T0292335

PRACTICAL PHYSICS

PRACTICAL PHYSICS

A Collection of Experiments
for Upper Forms of Schools & Colleges
together with the Relevant Theory

BY

SIR CYRIL ASHFORD

K.B.E., C.B., M.V.O., Hon.LL.D., M.A.

Sometime Assistant Demonstrator at the Cavendish Laboratory
Senior Science Master in Harrow School and
Headmaster of the Royal Naval Colleges
of Osborne and Dartmouth

CAMBRIDGE
AT THE UNIVERSITY PRESS
1950

CAMBRIDGE
UNIVERSITY PRESS

University Printing House, Cambridge CB2 8BS, United Kingdom

Cambridge University Press is part of the University of Cambridge.

It furthers the University's mission by disseminating knowledge in the pursuit of education, learning and research at the highest international levels of excellence.

www.cambridge.org
Information on this title: www.cambridge.org/9781107586284

First published 1950
First paperback edition 2015

A catalogue record for this publication is available from the British Library

ISBN 978-1-107-58628-4 Paperback

CONTENTS

PART II: LIGHT

PART III: HEAT AND ELECTRICITY

NOTE

The Local Examinations Syndicate was fortunate in securing Sir Cyril Ashford's help and interest after his retirement from Dartmouth in the preparation of practical physics exercises for the Higher School Certificate Examination. Most of the problems in this book were set in that examination between 1932 and 1947. They thus combine the originality and interest which they owe to their author with the merit of having been worked over by moderating examiners in the laboratory and at meetings. In their present form they have also benefited from the fact that the author was able to judge their suitability from the successful and unsuccessful solutions sent in by many examination candidates, and from the comments of their teachers. The Syndicate welcomes Sir Cyril Ashford's re-use of these practical problems as the basis of a teaching text-book which will make available in Schools and Universities the results of the author's pioneering efforts over a period of 15 years.

J. L. BRERETON

SYNDICATE BUILDINGS
CAMBRIDGE

February 1949

PREFACE

In the early stages of his science course a boy lacks reasoning power and first-hand acquaintance with the properties and behaviour of materials, so that he cannot fully comprehend a logically connected course of lecture-demonstrations. During those stages, laboratory work should therefore, so far as is practicable, go hand-in-hand with the lectures. By the time the boy enters what is commonly called the post-certificate stage the conditions have changed; he can then follow lectures and demonstrations without having previously handled the apparatus himself, and consciousness of the true relation between theory and practice in an inductive science is beginning to dawn on him. The need for individual experimentation is certainly not lessened at this stage, but its main purpose becomes for the time being the verification of theory which has been, or can be, developed by deductive logic from earlier, more fundamental, experiments, and the practical testing of hypotheses, sprung perhaps from his own, perhaps from a more mature, scientific intuition.

The emphasis should now be laid on the best way of using his available instrumental equipment to establish or reject a particular theoretical result, and on the extent to which his experiments do establish it. Laboratory technique and manipulative skill in general, the setting up of apparatus and the precautions for accuracy of observation in a particular case, and the reduction of those observations, all enter into this procedure. New powers of a general kind, applicable to any particular problem, have to be developed, so that there is no longer any strong reason for close relation between lecture and laboratory work; indeed, laboratory work may almost be kept in a separate compartment, to be developed concurrently with his growing body of theoretical knowledge, whereby each may be ready to come to the assistance of the other as need arises. Hence the exercises forming a laboratory text-book for this stage need not have any logical sequence and are therefore capable of being taken in any order (which is a great convenience in a school laboratory with its necessarily limited equipment), nor do they need to cover the whole of the lecture course.

There is already available a considerable body of traditional experiments designed to illustrate the lectures appropriate to this stage; to propose new ones calls for justification. Many of those included in this book, which is primarily designed for the last year of the school course, are intentionally concerned with theory which the schoolboy is unlikely to have studied, and the verification of such theory by experiment is a procedure in line with the foregoing arguments. In others the results to be tested can be deduced from theoretical work with which he is more or less familiar, but the general lay-out of the experiments and the methods of reducing the observations in such a way as to check the theory effectively may be unfamiliar to him. In every case the aim has been to present the problem as one in Practical Physics, with the laboratory functioning at least as the Court of Appeal rather than as a place where lectures can be revised at leisure, and preferably as a real laboratory where a boy can try out hypotheses and determine physical or instrumental constants, free from authoritarian influences.

The outlines of relevant theory with which experiments are prefaced in many text-books are often worded in such a way as to convey the impression that that theory settles the matter by its own authority, and that the boy succeeds in his experiments only in so far as his results agree with it; that he is to do the experiment for the sake of practice in experimenting, and that it is his manipulative skill, not the theory, which is being put to the test. It is in the belief that this is to deny Physics its true status as an inductive science that a vigorous attempt has been made here to banish any such impression from the boy's mind and to make him realise that teachers and text-books alike 'abide our question'.

Most of these exercises are based on questions set during the last 15 years in the Higher School Certificate papers of the Cambridge University Local Examinations Syndicate, and grateful acknowledgement is made to the Syndics for their permission to use them in this form. The exercises have of course been radically transformed and amplified to fit them to form part of a teaching course, and care has been taken to increase rather than diminish the extent to which their original design was influenced by the foregoing considerations. The use of examination questions as a basis for a teaching course has the merit that the problems set for solution in a practical examination must be so designed that they can be carried through in every one of a wide variety of laboratories, some at any rate possessed of a very slender

sequence, all the experiments in

exception, require no apparatus that does not already exist in every laboratory where the book is likely to be used. This puts into effect another general principle which the author believes to be of considerable importance, that a student should learn to rely on his own powers rather than on elaborate equipment; in C. V. Boys's words, not 'to bow down to the brazen image that the instrument maker has set up', but to make the very best use of what is to hand, however simple it may be.

In some of these practical problems the boy cannot be expected to know, to have been taught, or even to have access to books or publications containing the relevant theory. Notes on this theory have therefore been appended to individual exercises.

It is by no means essential, though it is desirable, for him to master the substance of these notes; they may in some cases be too advanced for him, but it will do him good to see how much of them he can follow at the stage he has reached in his theoretical work, and to realise that he can safely and properly explore by experiment ground which is beyond his present range of theory. From that point of view there is really no need for him even to glance at these notes, but they are at hand in case he is moved to do so. Their presence may be rather intimidating to a student unless he understands that this is their purpose, and that for him the actual experiment is the all-important matter; it is hoped that this will be made plain by the teacher and that he may not himself be misled into overrating the difficulty of the course. The fact that most of the experiments have been carried out under examination conditions by hundreds of VIth Form boys, with a measure of success represented by normal distribution curves, should dispose of such misapprehensions.

An Introduction has been provided in which a number of practical points in the reduction of observations is set forth. Teachers usually prefer that their pupils should follow local customs in the setting-up of apparatus and the choice of precautions for accuracy; they would be unlikely to welcome outside interference in these matters. On the other hand, the art of reducing observations may almost be said to be one and indivisible; it would take too long for a demonstrator to teach it separately to each pupil, but it is of such practical importance that the demonstrator may well wish to be able to refer the student to some treatise on the subject; such treatises are not usually to be found in elementary text-books; it therefore seemed desirable to attempt to meet the need here, even in this restricted form. It will be seen that graphical methods, with emphasis on linear equations, have been given the preponderance that is now common in teaching laboratories.

In the Introduction, appendices and some of the notes on theory a few unpublished pieces of theoretical work have been included; the author can only express the hope that they will withstand criticism and plead that he has at least offered his critics facilities for testing their truth by designing experiments for that purpose, in accordance with the principles set out in this preface.

INTRODUCTION

THE REDUCTION OF OBSERVATIONS

Not the least important part of a laboratory experiment is the utilisation of the observations made in the course of that experiment. The object of the experiment may be to measure some physical or instrumental constant, such as g or the focal length of a lens, or to find out how two or more variable quantities are related to one another, such as the volume, pressure and temperature of a constant mass of gas, or it may be a combination of the two purposes. We may have observed as accurately as we can the value of one of these variables corresponding to an observed value of another, and may have done so for several different values of the latter; the problem remains, how to use these pairs of measurements so as to deduce from them the algebraical relation between them.

The first part of this problem is to discover the *form* of this relation; the two columns of figures representing the crude results of our measurements, the 'pointer-readings' as Eddington called them, are often very unpromising material for this process. For instance, after the wave-lengths of the bright lines in the spectrum of hydrogen had been measured it needed a lot of co-operative effort before any mathematical law connecting them could be discovered; again, the positions of one or two planets at various known times had been fairly accurately measured for many years before Kepler could discover his three general, comparatively simple, laws governing the motion of all planets, from which Newton could deduce his supremely simple law of Universal Gravitation. These are instances of the application of genius to great problems; it may be a descent from the sublime to the ridiculous to point out that in Exp. 22 and again in Exps. 28, 29 and 30 problems of the same nature are set for your solution, but very humble illustrations may be the most illuminating.

The second part of the problem is to discover whether the law thus formulated is obeyed with complete precision throughout the whole range of the variables which, mathematically, it purports to cover. For instance, it was long thought to be established that light travelled to us from the stars in geometrically straight lines, but extremely precise measurements (undertaken because Einstein's work on relativity suggested that it might not be true in all circumstances) showed that a ray of light passing near the sun is deflected, very slightly but measurably, by the mass of the sun. Making again the descent from the sublime to the ridiculous, we shall presently have occasion to consider the way in which a spiral steel spring fails in practice to obey Hooke's law.

There is therefore scope for great ingenuity in reducing tables of observations to a form that will be helpful in each of the foregoing processes. For the first, there is clearly need for more than ingenuity; scientific imagination or intuition of the highest order may be required to suggest the form of the relation. But when it is merely a question of testing the truth or applicability of the relation, of 'verifying' it, a more lowly intelligence may suffice, and it may become merely a matter of choosing the most suitable out of a number of well-recognised manipulations of the crude results of the observations. For example, the mere addition or subtraction of a suitable constant in each case (as in Exp. 22 mentioned above) may make it easy to see the next step in evolving the required law; or it may be useful to take the reciprocal, or the logarithm, of each, and so on.

For this reason the following notes on some of the recognised methods of reduction and presentation may be found helpful.

1. *Computation of a Single Constant*

(*a*) Laboratory work often consists of determining, as accurately as we can, the value of a single physical or instrumental constant, assuming the truth of a relation established by theoretical reasoning. For example, we may have to determine the most probable value of the instrumental constant f of a particular lens, accepting the truth of the general lens formula $1/u + 1/v = 1/f$, or we may have to find by experiments on a simple pendulum the value of the physical constant g, accepting the truth of the theoretical relation $T = 2\pi\sqrt{(l/g)}$.

Taking the former example, we set up the apparatus and measure, for one setting of the object, the values of u and v. We substitute these values in the general relation and compute the value of f. If we repeat this operation with a different setting of the object it is almost a certainty that we shall obtain a different value of f; suppose we do this N times in all, we are left with N different values of f, and are faced with the problem of deducing from them a value which is most likely to be the true value of f. Denote this true value by F, and the various calculated values by f_1, f_2, f_3, etc. Then the error of the first determination was $f_1 - F$, of the second was $f_2 - F$, and so on, and the sum-total of all the errors was $\Sigma f_1 - NF$, where Σf_1 means the sum of all terms like f_1.

Suppose that it is equally probable that any one of the calculated values is correct, and that their number N is infinitely large; then the Theory of Probability shows that for every one that is too large by a certain amount we can find one that is too small by exactly the same amount, and the arithmetic mean of these two will be exactly equal to F, and the sum-total of their errors is zero. Repeating this for all the observations, we see that $\Sigma f_1 - NF$ must be zero, or $F = \Sigma f_1/N$, or F, the true value, is the arithmetic mean of all the calculated values.

We cannot, of course, take an infinite number of observations, but reasons are given in §4 for taking the arithmetic mean of such calculated values as are available as the most probable value of F.

It may happen that one, or more, of these observations leads to a value outstandingly different from the average value; it is tempting to ignore it out of hand as incorrect. It is clearly sound policy to go back and repeat this observation with great care, to find out whether there has been an error of observation or calculation. If there was, we can obviously ignore the original observation and substitute the revised one; if not, it is on the whole the soundest policy to retain it.

(*b*) This process of computing the arithmetic mean is in many cases quicker than a graphical method of determining the constant; it can be carried to a higher degree of numerical accuracy, and it has the additional merit of leading to a numerical estimate of the degree to which our experiments are trustworthy.

Take, for instance, the numbers in the first and second columns of this table, which are the reciprocals of the values of u and v found in the course of Exp. 25 (B) with a concave lens:

$1/u$	$1/v$	$-1/f$	$-f$	Difference from mean
0·01117	0·07681	0·06564	15·23	0·00
0·01447	0·07987	0·06540	15·29	+0·06
0·02086	0·08681	0·06605	15·14	−0·09
0·02574	0·09076	0·06504	15·37	+0·14
0·02995	0·09506	0·06611	15·13	−0·10
0·03402	0·09982	0·06580	15·20	−0·03
			6|91·36	6|0·42
			15·23	**0·07**

The third column shows the difference between the first and second, which in accordance with the accepted theory is $-1/f$; the fourth shows $-f$ (taken from a table of reciprocals) and the way it leads to the arithmetic mean value for f of 15·23. This is the best value of f which we can get from our observations.

We can consider the fifth column as giving the 'error' of each observation; if we take the arithmetic mean of these numbers, disregarding their signs (as shown), we can consider the result, 0·07, as representing the mean error. It is customary to express these results in the form $f = -15{\cdot}23 \pm 0{\cdot}07$.

Since 0·07 is about $\frac{1}{2}$% of the value of f, it is often said that these experiments are 'accurate to $\frac{1}{2}$%'. This shorthand statement is convenient for purposes of comparison but liable to mislead unless we bear in mind what it really means; for instance, one of these experiments shows an error of almost 1 % of the arithmetic mean of the values we have obtained for f.

2. *Graphical Representation; Linear Graphs*

Laboratory work usually involves more complex problems than those dealt with in § 1, where we assumed that the relation between the measured variables was known; the problem is often to determine by experiment what the relation is. Let us assume for the present that our measurements are absolutely accurate; we can afterwards consider how best to allow for observational errors.

Suppose that we plot in the usual way the points represented by the pairs of observations, using such scales along the two axes as will spread the points satisfactorily on the graph paper. Suppose that we find that these points lie on a straight line; the relation between the observed quantities is clearly linear, of the form $ax + by = 1$, where x and y are the quantities and a and b are some constants.

If this graph is not a straight line, let us suppose that by some manipulation of the observed quantities (e.g. squaring one or both or taking logs of one or both), and plotting these manipulated quantities we can get a straight-line graph. Then, again, the form of the relation is discovered.

Further, a and b are at once measurable since the intercept on the axis of x, found by putting $y = 0$ in the equation $ax + by = 1$ to be $1/a$, can be read off as the value of the intersection of the straight-line graph and the axis of x in terms of the scale for that axis. Similarly,

$1/b$ can be found. But it is essential to be on one's guard in carrying out even this familiar procedure.

(*a*) Suppose that a closely coiled spiral steel spring is hung up and loaded with a series of weights (W g.) ranging from 0 to 1300 g., and the corresponding overall lengths (l cm.) of the spring are measured; the points representing l and W are plotted and the full-line graph in Fig. 1 drawn. Let us assume that the graph between A and B is, as it appears to be, a straight line.

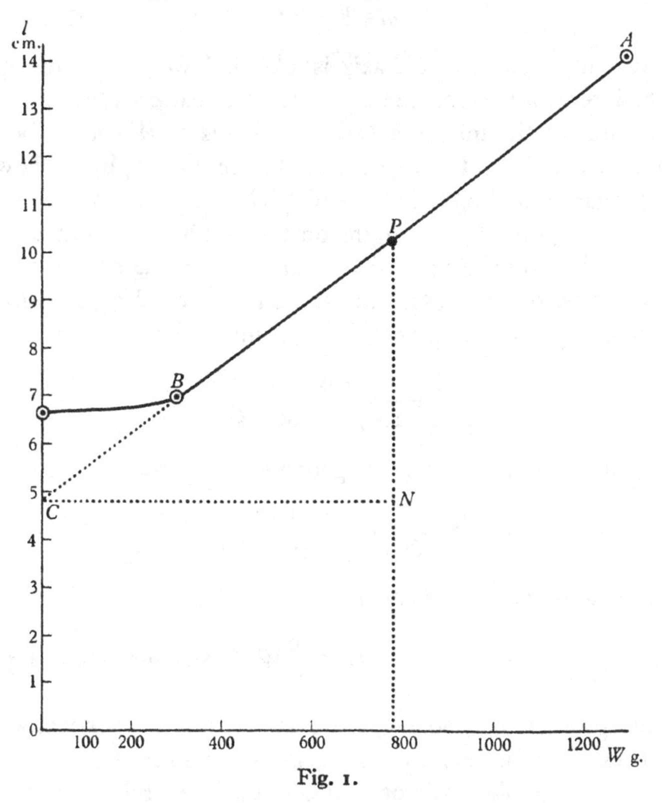

Fig. 1.

It is easy to account for the horizontal part of the graph, since, let us suppose, the experimental set-up showed that at a load of 300 g. consecutive turns of the wire were in contact, and the spring could not shorten further as the load was reduced.

Hence any linear algebraical relation connecting l and W which we may find to be true for loads on this spring between 300 and 1300 g. may be true for greater loads, but it certainly cannot be true for loads less than 300 g.

Nevertheless, any relation between the co-ordinates l and W which holds good for a portion of a straight line must hold good mathematically for the whole of it, produced to infinity both ways. Hence in finding the algebraical relation which holds good physically only for the part AB of the graph we may produce that part geometrically to cut the axis of l at C (as shown by the dotted line) and beyond C to cut the axis of W at D. Hence if the required relation between l and W is

$$al+bW=1,$$

since the intercept on the l-axis is $1/a$, and we see by the graph that the intercept is 4·80, we deduce at once that $a=1/4{\cdot}80$.

The value of the intercept OD on the axis of W cannot be read off so immediately, but can be deduced from the graph as follows.

By similar triangles, $OD/OC=NC/NP$. Here we are assuming that these are geometrical lengths on the graph paper, but the proportion holds good if we replace OD and CN by the quantities of W represented by those lengths, and OC and NP by the quantities of l represented by those lengths. Hence, numerically,

$$\frac{OD}{4{\cdot}80}=\frac{780}{10{\cdot}30-4{\cdot}80}=\frac{780}{5{\cdot}50},$$

so that, since OD represents a negative value of W,

$$b=\frac{1}{OD}=-\frac{5{\cdot}50}{4{\cdot}80\times 780},$$

and the relation between l and W is

$$\frac{l}{4{\cdot}80}=\frac{5{\cdot}50}{4{\cdot}80\times 780}W+1, \quad \text{or} \quad l=\frac{5{\cdot}50}{780}W+4{\cdot}80=\mathbf{0{\cdot}00705\,W+4{\cdot}80},$$

with the proviso that, so far as we know from our experiment, it holds good only for values of W between 300 and 1300.

This is the simplest way of setting out a linear relation. But it is sometimes convenient to write the above relation in the form

$$l=4{\cdot}80(1+0{\cdot}001465W),$$

which is a particular case of the general form $y=y_0(1+\alpha x)$, where y_0 is the value of y when $x=0$. The coefficient of W (that is, 0·001465) is then a constant for all springs of that particular kind, whatever their length may be, and it would be reasonable to call this constant the coefficient of elasticity of such springs.

Note particularly that in this expression for l, y_0 is 4·80, or the intercept of the straight line on the axis of l and not the intercept of the physical graph on that axis, which is 6·60 and represents the actual length of the spring when $W=0$. We may, in fact, call y_0, the value of y when $x=0$, the 'ideal unstretched length', if the relation $y=y_0(1+\alpha x)$ held good when $x=0$ (which it does not).

This has been set down at, perhaps, tedious length, since confusion can easily and often does arise on the point. Hooke's law holds good in this spring only when W is more than 300, and beyond that point the extra extension produced by any additional load is proportional to that additional load, but the total extension is never proportional to the total load. However, the total length l of the spring for any particular value of W can be read off the graph directly, without reference to unstretched lengths and unloaded springs; conversely, of course, the load on the spring can be deduced directly from the graph when the total length l is known.

(*b*) Next, suppose that the periodic times (T sec.) of simple pendulums of various lengths (l cm.) have been measured, producing the first two columns of this table:

l cm.	T sec.	T^2
60	1·55	2·40
80	1·79	3·20
100	2·00	4·00
120	2·19	4·80

Plot the points corresponding to these values of l and T, as (i) in Fig. 2, and draw the best-fit smooth graph among them. This line is so slightly curved that it would be almost excusable to regard it as straight. If we deduce, as in the foregoing example, the relation between T and l we should find it to be

$$T=0·01025l+0·93.$$

But the above values of l are badly chosen; they do not cover more than a small fraction of the range that can easily be dealt with in the experiment. If we take small values of l, such as 16 and 9, we shall get 0·80 and 0·60 for T, giving the plotted points P and Q. These obviously do not fall on the straight line whose equation has been found; that relation must, therefore, be wrong, and the true relation cannot be linear, of the form $T=al+b$.

If we draw a smooth curve through all these plotted points it is not very illuminating; it is part of a curve which may well pass through the origin, and it looks like a parabola, but that does not enable us to determine at once the relation between l and T, as we could do when the graph was linear.

But the theory of simple harmonic motion, or the practice of Exp. 4, shows that the relation should be $T^2 = Al$, where A is a constant; if then we plot l against T^2 instead of T, we should theoretically get a linear graph, passing through the origin. Filling in the

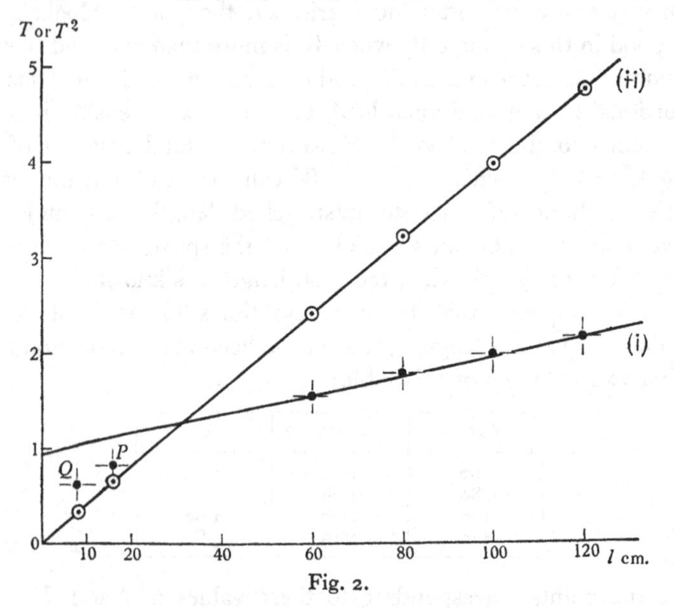

Fig. 2.

third column of the table and taking the squares of 0·80 and 0·60 to correspond with 16·0 and 9·0 for l, we get (ii) in Fig. 2. This may be regarded as confirming the theory by being a straight line through the origin; its slope to the l-axis can be read off the graph as 4 in 100, so the experimental relation is $T^2/l = \frac{4}{100}$, or

$$\mathbf{T^2 = 0{\cdot}04l.}$$

This illustrates the advantages to be gained by manipulating one or both of the measured quantities so that when plotted the corresponding points will, or should, lie on a straight line. To know what sort of manipulation is needed we must know the general form of the relation; theory often furnishes us with this information, as it did in the above instance. But a certain amount of intelligence is often required in addition to theoretical knowledge, for no general rules can be given by which we can dispense with that intelligence.

(*c*) But a good many particular cases fall into one or other of a few categories. Let us denote the experimentally measured quantities by x and y, and unknown constants by a, b, c, etc. The above case of a simple pendulum is a particular case of the general form $\mathbf{y^n = ax + b}$, n in that case being 2 and b being 0. In the general case we must plot y^n against x to get the linear graph from which we determine a and b. Other cases of the same class are given by Boyle's law and Newton's lens formula, and the relation between frequency and length in a monochord; here $b = 0$ and $n = -1$, so that we must plot the reciprocal of y against x.

(*d*) Many problems in Optics, and some which are concerned with resistances in parallel, involve the relation $1/x + 1/y = \text{const.}$, and this gives a linear graph if we plot the reciprocals of both x and y.

(*e*) Another group, including the temperature of a hot body losing heat to its colder surroundings, the charge in a condenser losing its charge through a resistance, and the speed of a body coming to rest because of 'frictional' resistance proportional to the speed (as in a damped ammeter), involve the relation $y = y_0 e^{-ax}$, where x is the elapsed time and y_0 is the value of y when $x = 0$, and $e = 2{\cdot}718\ldots$. Taking logs of both sides we get $\log y = \log y_0 - ax \log e$; since a, $\log e$ and $\log y_0$ are constants, we get a linear graph if we plot $\log y$ against x.

(*f*) Confusion sometimes arises between (A) Hooke's law, Charles's law, etc., where the general relation is $y = y_0(1 + ax)$, (B) thermal expansions of solids and liquids, with the same general relation, and (C) the group of cases just mentioned; it may be as well to clear up the position to some extent.

(A) When the change in y may be considerable, as in the thermal expansion of gases under constant pressure, y_1 and y_2 may differ considerably from y_0, and y_0 must therefore be precisely defined.

(B) When the change in y is small in comparison with y, as in the thermal expansion of a solid or liquid, it makes little difference whether we represent the expansion as $y_0 a$, $y_1 a$ or $y_2 a$ times the change of temperature. This vagueness is harmful only if it leads to the belief that the rate of change of length or volume of the body with temperature is precisely proportional to that length or volume, in which case it should, strictly speaking, come under the next heading (C).

(C) Consider a condenser of capacity C farads, charged with a quantity Q coulombs of electricity which raises the P.D. between the plates to V volts; let the plates be connected through a high resistance R ohms, then the current (i amp.) through the resistance will have a value of V/R or Q/CR, at any time (t) when the charge is Q. Since the current equals the rate at which electricity passes through the resistance, i also equals the time-rate of decrease of Q. If then we plot a graph connecting Q with t, the current at any time will be measured by the slope of the tangent to the graph at the point representing that time. Hence the graph must be of such a form that the slope of the tangent at any point is directly proportional to the ordinate of that point. Mathematics shows (see note to Exp. 16) that in that case the relation between points on the curve must be $Q=Q_0e^{-at}$, where Q_0 is the value of Q when $t=0$. From this relation we can get a linear graph, as shown above, in (a).

(g) The experiments in this book furnish many examples of the manipulation of the observed quantities to give linear graphs; it will be seen that in a considerable proportion of them there comes a stage when the results of theoretical reasoning are quoted, expressed as an equation which has already been transformed mathematically so as to be effectively linear (although it may contain powers of the variables higher than the first, or logs of them, etc.). In effect, these transformations may be regarded as part of the laboratory work and not of the theory, which does not itself call for such transformations.

This equation is useful in the practical work in two ways: first, because it may point to the way in which the observations must be manipulated to produce a linear graph so that we may check the *form* of the law, or the relation between the variables, by testing the straightness of the experimental graph; secondly, because we can get accurate numerical values of the two constants a and b in the equation $ax+by=1$ representing that law, by measuring or calculating the intercepts on the two axes. The theoretical equation gives the values of a and b in terms of the magnitudes of various quantities in the apparatus used in the experiment, which are unchanged throughout that experiment, and we can check the theory in this respect by substituting the measured values of these quantities in our experiment in the theoretical expressions for a and b, and comparing the results with our measured values of a and b.

3. *Distorted Graphs*

If we have measured in a laboratory several pairs of numerical values of two connected physical quantities, each expressed in its appropriate units, A and B say, and wish to represent the results graphically, the simplest way to do so is to use a sheet of squared paper, divided into inches (say) and tenths, and to take an inch horizontally to represent A and an inch vertically to represent B, and number each axis progressively from the origin. The point (such as P_1) corresponding to each observation is then plotted as usual, and a smooth curve drawn to fit these points as nearly as possible; it is assumed in Fig. 3*a* to be a straight line. Let us call such a curve the ***true*** graph of the experiment; it is what a mathematician means by the phrase 'the curve represented by an equation'.

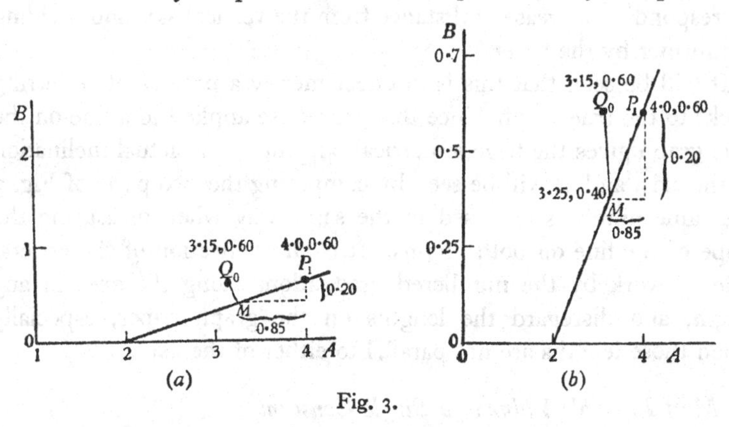

Fig. 3.

This graph may be badly spaced on the paper; we usually rectify this by changing the scales on one or both of the axes, making an inch on the axis represent many times, or a fraction of, A or B; the two axes are then numbered off according to these new scales, as in Fig. 3*b*, and the points are plotted and the smooth curve drawn exactly as before.

What we have done in effect is to stretch, or shrink, the paper carrying the graphs and all lines connected with it; the stretching or shrinking takes place in the direction of the axes, and is usually different in the two directions. Comparison of the two parts of Fig. 3 shows that this procedure produces a geometrical distortion of the true graph; let us then call the result the ***distorted*** graph. In particular, lines such as Q_0M and P_1M which are at right angles in the true graph are far from perpendicular to one another in the distorted graph. A tangent at a point of a curve in the true graph is still a tangent in the distorted graph, but the normal which is at right angles to that tangent in the true graph looks absurd in the distorted graph; a circle distorts into an ellipse, and so on.

The only way in which the beginner is likely to be troubled by the distortion is in the matter of measuring the inclination of a straight line to one of the axes; that inclination is obviously increased by any stretching perpendicular to that axis. He is therefore taught not to measure the angle of inclination by a protractor and take the tangent of that angle, either on the distorted or true graph, but to measure what he learns to call the 'slope' of the line, by finding by means of the graduations on the vertical axis the increase, for two points on the line, of the distance from the horizontal axis, and the corresponding increase of distance from the vertical axis and dividing the former by the latter.

It will be seen that this is in effect merely a process of 'referring back' to the true graph, since this procedure applied to a line on the true graph gives the trigonometrical tangent of the actual inclination to the axis, and as will be seen by comparing the two parts of Fig. 3 the same numbers are used in the same way when measuring the slope of the line on both graphs. It is an application of the general rule to work by the numbered graduations along the axes in any graph, and disregard the lengths on the graph paper, especially when those lengths are not parallel to either of the axes.

4. *Most Probable Value of a Single Constant*

The problem of deducing the most probable final result, whether that be a single constant or a linear graph, from the results of a few experiments when those results are not all identical was shirked in the earlier parts of this Introduction; it is so important that it must now be dealt with.

The general treatment will be much easier to follow if we first analyse in detail a simple concrete case. Suppose that we have made three sets of measurements of a single quantity such as the focal length of a certain lens, and have got 20·60, 20·20 and 20·10 cm. Let us denote by x_0 the most probable value of the quantity, 'most probable value' meaning the closest approximation to the *true* value that we can deduce from our observations.

These observations may be incorrect for either or both of two reasons. First, the apparatus may be faulty; for instance, it may contain a paper scale of lengths which expands when damp, or there may be an undetermined zero error; errors on this account are often termed 'systematic errors' and strictly speaking are not covered by this section. Secondly, the setting of a moveable component or the reading taken may be faulty, for example in estimating tenths of a scale division; these are usually termed 'casual errors of experiment' and will for brevity be called 'errors' in this section.

The 'error' of any observation is properly the difference between the observed and the true values; since we can never know the absolutely true value we will take the error to be the amount by which the observed value exceeds x_0, the most probable value deduced in some way from our observations. Hence the error may be positive or negative.

It is reasonable to assume that the observed values are on the whole grouped closely round the most probable value x_0.

(*a*) It would be reasonable to choose x_0 so that it makes the aggregate of all the errors, that is, their numerical sum without regard to their signs, as small as possible. Now in this case, for such close grouping, x_0 must lie somewhere between the largest, 20·60, and the smallest, 20·10, and the numerical values of the errors of these two observations will be $20{\cdot}60 - x_0$ and $x_0 - 20{\cdot}10$. Hence whatever value x_0 may have between 20·60 and 20·10, their numerical sum will be $20{\cdot}60 - x_0 + x_0 - 20{\cdot}10$ or **0·50**. To this must be added, in order to get the aggregate, the numerical value of the second error, which is $20{\cdot}20 \sim x_0$, without regard to sign; whatever its sign, this addition will be smallest when x_0 equals 20·20, for this error is then zero. Hence on this assumption **20·20** is the most probable value of x_0, and the numerical sum of the errors is $0{\cdot}40 + 0{\cdot}00 + 0{\cdot}10$ or **0·50**.

(*b*) Another possible assumption for getting x_0 is that the *algebraic* sum of the errors should be zero.

This leads to $20{\cdot}60 - x_0 + 20{\cdot}20 - x_0 + 20{\cdot}10 - x_0 = 0$ or

$$x_0 = \tfrac{1}{3} \times 60{\cdot}90 = \mathbf{20{\cdot}30}.$$

This is obviously equivalent to adopting the arithmetic mean of the observed values as the most probable value, which is the usual custom. It brings out a different value for x_0, 20·30 instead of 20·20 as in (*a*); the numerical sum of the errors is now 0·6 instead of 0·5.

(*c*) A third possible assumption avoids all trouble with signs, by dealing with squares, which are always positive. It is, that the sum of the squares of the errors should be a minimum.

Let us try, without using anything but simple arithmetic, how well the above values of x_0 (20·20 and 20·30) agree with this assumption. If we calculate the sums of the squares of the errors when x_0

is 20·195, 20·200 and 20·205 we find them to be 0·1730, 0·1700 and 0·1670 respectively. So the sum decreases continuously, and is not a minimum when $x_0 = 20{\cdot}20$. Repeating this for $x_0 = 20{\cdot}30$, the sums of the squares when x_0 is 20·295, 20·300 and 20·305 are 0·14009, 0·14000 and 0·14009 respectively, showing a minimum at $x_0 = 20{\cdot}30$.

Hence this assumption leads to the same results as the Arithmetic Mean method of (*b*), but not as (*a*).

(*d*) Each of these three assumptions is reasonable; how are we to choose between them? They have the common merit that, as shown in the Theory of Probability, they all give identically the same result when applied to an infinitely large number of observations; but in practice we are concerned only with a small number of observations. However, there is one distinct difference between the assumptions; the first does not lend itself at all to mathematical treatment, the second does so, and the third does so more widely and in some cases (one of which we shall soon come across) it is the only one that can be used. This third one is called the Method of Least Squares.

Hence, when dealing with a small number of observations of a single constant we will adopt the result of the second and third assumptions, that *the most probable value is the arithmetic mean of the observed values.*

(*e*) Replacing the particular case of (*b*) and (*c*) by a more general treatment, if we denote by ΣE_1 the algebraic sum of the errors in all the N observations which gave observed values x_1, x_2, x_3, etc., the assumption in (*b*) gives

$$0 = \Sigma E_1 = \Sigma(x_1 - x_0) = \Sigma x_1 - \Sigma x_0 = \Sigma x_1 - N x_0.$$

But $\dfrac{\Sigma x_1}{N}$ is the arithmetic mean of x_1, x_2, etc., so $x_0 = \dfrac{\Sigma x_1}{N} =$ the arithmetic mean of the observed values.

Again, the assumption in (*c*) is that $(x_1 - x_0)^2 + (x_2 - x_0)^2 +$ etc. is a minimum for changes in x_0. Books on the differential calculus show that this happens when $\dfrac{d}{dx_0}\{(x_1 - x_0)^2 + (x_2 - x_0)^2 + \text{ etc.}\} = 0$,

or when
$$-2(x_1 - x_0) - 2(x_2 - x_0) - \text{ etc.} = 0,$$

or when
$$-2\{\Sigma x_1 - N x_0\} = 0,$$

or when
$$x_0 = \frac{\Sigma x_1}{N} = \text{arithmetic mean of the observed values as before.}$$

(5) *Most Probable Linear Equation, or best-fit straight line*

It may perhaps be advisable to preface this section with the words inscribed over the entrance to Plato's Academy, ἀγεωμέτρητος μηδεὶς εἰσίτω, although it is the quantity rather than the advanced quality of the mathematics in it that may deter him.

If we want to find the most probable position of the straight-line graph among a number of plotted points representing observations, which we may call the 'best-fit line', we have a less simple task, for this line may be displaced broadside, or may turn about a fixed point, or both simultaneously.

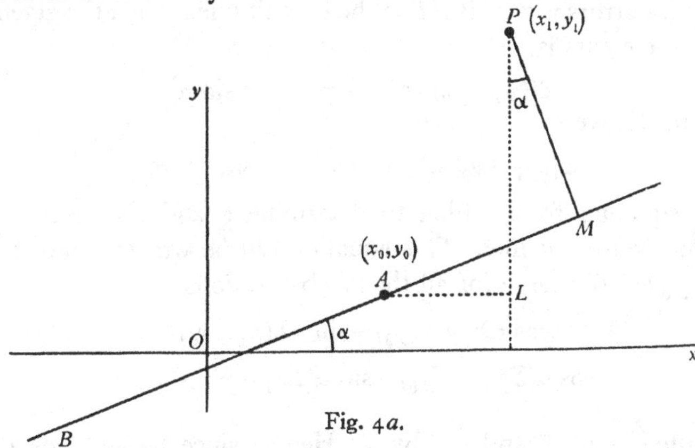

Fig. 4*a*.

Let us make the reasonable assumption that the error of any of the observations can be measured by the length of the perpendicular, from the point *P* on the graph paper representing the observation, dropped on the best-fit line.

Hence in Fig. 4*a*, if *AB* is the best-fit line, the error of observation *P* is *PM*.

Let x_1 and y_1, x_2 and y_2, etc. denote the pairs of observations, represented on the graph by points P_1, P_2, etc.; we want to find the best-fit straight line among these points, or the linear equation which is most nearly satisfied by x_1 and y_1, etc. It should be noted that these values of x_1, x_2, etc. probably differ widely in magnitude and perhaps in sign, unlike the x_1, x_2, etc. of § 4 (*c*) which are all nearly equal in magnitude and almost certainly of the same sign. Hence Σx_1 represents the algebraic sum of x_1, etc., each keeping its appropriate sign; $\frac{\Sigma x_1}{N}$ may therefore differ widely from any particular value of x.

Denote the co-ordinates of some point A on the line by x_0, y_0; then if AL and PL are parallel to the axis of x and y, $AL = x_1 - x_0$ and $PL = y_1 - y_0$. Suppose that AB makes an angle α with the axis of x; this is the inclination of the graph when the scales along the two axes are equal, not the actual angle on a distorted graph (see §3). From Fig. 4*a* we see that the error

$$\begin{aligned} PM &= PL \cos\alpha - AL \sin\alpha \\ &= (y_1 - y_0)\cos\alpha - (x_1 - x_0)\sin\alpha. \qquad (1) \end{aligned}$$

(*a*) Let us first apply condition (*b*) of §4, that the algebraic sum of all these errors is zero if AB is the best-fit line. Then the algebraic sum of the errors is

$$\Sigma\{(y_1 - y_0)\cos\alpha - (x_1 - x_0)\sin\alpha\}$$

so we must have

$$\Sigma\{(y_1 - y_0)\cos\alpha - (x_1 - x_0)\sin\alpha\} = 0 \qquad (2)$$

as the equation from which to determine α and the fixed point x_0, y_0 on the best-fit line. This equation can be written, since α and x_0 and y_0 are the same for all the N observations,

$$\cos\alpha\Sigma(y_1 - y_0) - \sin\alpha\Sigma(x_1 - x_0) = 0$$

or

$$\cos\alpha(\Sigma y_1 - Ny_0) - \sin\alpha(\Sigma x_1 - Nx_0) = 0. \qquad (3)$$

Denote $\frac{\Sigma x_1}{N}$ by $\bar{x}$, and $\frac{\Sigma y_1}{N}$ by $\bar{y}$. Hence, since neither $\cos\alpha$ nor $\sin\alpha$ can be infinitely large, this equation is satisfied if $y_0 = \bar{y}$ and $x_0 = \bar{x}$. Hence the best-fit line must pass through the point $\bar{x}, \bar{y}$.

The equation (3) is satisfied by these values of x_0 and y_0, whatever be the value of α; hence condition (*b*) of §4 is satisfied by any straight line through $\bar{x}, \bar{y}$, and it does not suffice to fix completely the best-fit line. We must therefore try another, more effective, condition; let us try the Method of Least Squares, the condition (*c*) of § 4.

(*b*) This condition is that ΣPM^2 must be a minimum when AB is the best-fit line.

Now

$$\begin{aligned} PM^2 &= \{(y_1 - y_0)\cos\alpha - (x_1 - x_0)\sin\alpha\}^2 \\ &= (y_1 - y_0)^2\cos^2\alpha + (x_1 - x_0)^2\sin^2\alpha - 2(y_1 - y_0)(x_1 - x_1)\sin\alpha\cos\alpha. \end{aligned}$$

Hence
$$\Sigma PM^2 = \Sigma(y_1 - y_0)^2\cos^2\alpha + \Sigma(x_1 - x_0)^2\sin^2\alpha - 2\Sigma(y_1 - y_0)(x_1 - x_0)\sin\alpha\cos\alpha.$$

Denote this expression by u for brevity.

Books on the differential calculus show that u is a maximum or a minimum for changing values of x_0 and y_0 and α when

$$\frac{du}{dx_0}=0,\quad \frac{du}{dy_0}=0 \quad\text{and}\quad \frac{du}{d\alpha}=0$$

respectively.

Now for a constant value of α

$$u=\cos^2\alpha\Sigma(y_1-y_0)^2+\sin^2\alpha\Sigma(x_1-x_0)^2-2\sin\alpha\cos\alpha\Sigma(y_1-y_0)(x_1-x_0). \tag{4}$$

Then if y_0 alone varies

$$\begin{aligned}\frac{du}{dy_0}&=-2\cos^2\alpha\Sigma(y_1-y_0)+2\sin\alpha\cos\alpha\Sigma(x_1-x_0)\\ &=-2\cos^2\alpha(\Sigma y_1-\Sigma y_0)+2\sin\alpha\cos\alpha(\Sigma x_1-\Sigma x_0).\end{aligned}$$

We have denoted Σy_1 by $N\bar{y}$ and Σx_1 by $N\bar{x}$, and since x_0 and y_0, whatever they may be, are the same for all the N observations,

$$\Sigma y_0=Ny_0 \quad\text{and}\quad \Sigma x_0=Nx_0.$$

Hence

$$\frac{du}{dy_0}=-2\cos^2\alpha(N\bar{y}-Ny_0)+2\sin\alpha\cos\alpha(N\bar{x}-Nx_0). \tag{5}$$

Similarly

$$\frac{du}{dx_0}=-2\sin^2\alpha(N\bar{x}-Nx_0)+2\sin\alpha\cos\alpha(N\bar{y}-Ny_0). \tag{6}$$

Both (5) and (6) vanish if $y_0=\bar{y}$ and $x_0=\bar{x}$. ΣPM^2 will therefore be a maximum or minimum for any particular value of α if the line AB passes through $\bar{x}$, $\bar{y}$. This is what we found before, but we have now a further condition in hand by which we may determine α.

In using this condition we can substitute $\bar{x}$ for x_0 and $\bar{y}$ for y_0 in (4). Then if α varies, (4) gives

$$\begin{aligned}\frac{du}{d\alpha}=&-2\cos\alpha\sin\alpha\Sigma(y_1-\bar{y})^2+2\sin\alpha\cos\alpha\Sigma(x_1-\bar{x})^2\\ &-2(\cos^2\alpha-\sin^2\alpha)\Sigma(y_1-\bar{y})(x_1-\bar{x}).\end{aligned} \tag{7}$$

Hence the condition determining α for a maximum or minimum of ΣPM^2 is that (7) equals zero. Dividing by $-2\cos^2\alpha$ and putting $m=\tan\alpha$, we must have

$$0=m\{\Sigma(y_1-\bar{y})^2-\Sigma(x_1-\bar{x})^2\}+(1-m^2)\Sigma(y_1-\bar{y})(x_1-\bar{x}).$$

Solving this quadratic,

$$m=A\pm\sqrt{(A^2+1)},\quad\text{where}\quad A=\frac{\Sigma(y_1-\bar{y})^2-\Sigma(x_1-\bar{x})^2}{2\Sigma(y_1-\bar{y})(x_1-\bar{x})}. \tag{8}$$

Hence the straight line which we seek passes through the point $\bar{x}, \bar{y}$ and is inclined to the axis of x at an angle α whose tangent m is given by (8).

But (8) gives two possible values for m; denote them by m_1 and m_2. Then the equation of the best-fit line is either

$$y-\bar{y}=m_1(x-\bar{x}) \tag{9}$$

or

$$y-\bar{y}=m_2(x-\bar{x}). \tag{10}$$

It is easy to determine which of these two is the best-fit line, by substituting a pair of observed values, say $x_1 y_1$, which is far from $\bar{x}, \bar{y}$, in (9) and (10). Only one equation will be nearly satisfied, and this contains the value of m to be adopted in the equation to the best-fit line.

It is also easy to see what the other equation represents; from (8)

$$m_1 m_2=\{A+\sqrt{(A^2+1)}\}\{A-\sqrt{(A^2+1)}\}=A^2-(A^2+1)=-1.$$

This shows that the straight lines represented by (9) and (10) are at right angles to one another; in fact, if one is the best-fit line, the other is the worst-fit line, corresponding to the condition that the sum of the squares of the perpendiculars on it is a *maximum*.

(*c*) A worked-out numerical example will help to make this general theoretical reasoning more clear, and it may be a sufficient guide to those who wish to make use of the method without understanding the whole or any part of the reasoning on which the method is based.

Suppose that we have obtained the five pairs of values x and y shown in the first two columns of this table, either by direct observation or by manipulating direct observations in such a way as to get a linear graph; some poor observations have been purposely included, as will be seen in Fig. 4*b*.

x	y	$x-\bar{x}$	$(x-\bar{x})^2$	$y-\bar{y}$	$(y-\bar{y})^2$	$(x-\bar{x})(y-\bar{y})$ +	$(x-\bar{x})(y-\bar{y})$ −
7·30	3·40	−2·330	5·429	−6·302	39·72	14·68	
8·23	7·00	−1·400	1·960	−2·702	7·301	3·782	
9·33	11·91	−0·300	0·090	2·208	4·875	—	0·662
11·42	10·89	1·790	3·204	1·188	1·411	2·127	
11·87	15·31	2·240	5·018	5·608	31·45	12·56	
5 | 48·15	5 | 48·51		15·701		84·757	33·149	
						0·662	
9·630	9·702					32·487	
$\bar{x}$	$\bar{y}$		$\Sigma(x_1-\bar{x})^2$		$\Sigma(y_1-\bar{y})^2$	$\Sigma(x_1-\bar{x})(y_1-\bar{y})$	

In this table $\bar{x}$ and $\bar{y}$ are found as the arithmetic means of the numbers in each of the first two columns in the usual way. The third column contains the values of $x-\bar{x}$ with the appropriate signs, and the fourth the values of $(x-\bar{x})^2$, and consequently of $\Sigma(x_1-\bar{x})^2$, the sum of all the terms like $(x_1-\bar{x})^2$.

In the fifth and sixth, the same is done for $\Sigma(y_1-y)^2$; the seventh and eighth give the products of the third and fourth, paying due regard to signs; these give the algebraical sum $\Sigma(x_1-\bar{x})(y_1-\bar{y})$.

In this case A in (8) becomes

$$\frac{84{\cdot}757-15{\cdot}701}{2\times 32{\cdot}487} \text{ or } 1{\cdot}063, \text{ and therefore } A^2=1{\cdot}130.$$

Substituting these in $m=A\pm\sqrt{(A^2+1)}$ we get

$$m=1{\cdot}063\pm\sqrt{(2{\cdot}130)}=1{\cdot}063\pm 1{\cdot}459$$

$$=2{\cdot}522 \text{ or } -0{\cdot}396. \qquad (11)$$

Hence from the theoretical reasoning in § 5(*b*) the best-fit line must pass through the point $\bar{x}$, $\bar{y}$ or 9·630, 9·702 and must have a slope to the axis of x of either 2·522 or −0·396. That is, the relation between x and y in the best-fit line must be either

$$y-9{\cdot}702=2{\cdot}522(x-9{\cdot}630) \qquad (12)$$

or

$$y-9{\cdot}702=-0{\cdot}396(x-9{\cdot}630). \qquad (13)$$

To find out which of these it is, substitute a pair of observed values of x and y from the table for x and y in each of these equations. Thus

$$3{\cdot}40-9{\cdot}702=2{\cdot}522(7{\cdot}30-9{\cdot}630) \quad \text{or} \quad -6{\cdot}302=-2{\cdot}522\times 2{\cdot}330,$$

$$3{\cdot}40-9{\cdot}702=-0{\cdot}396(7{\cdot}30-9{\cdot}630) \quad \text{or} \quad -6{\cdot}302=+0{\cdot}396\times 2{\cdot}330.$$

It is obvious that the first equation is nearly satisfied, and the second is not, so it is (12) that is the 'most probable relation' between x and y, deduced from the observations.

It may be pointed out at this point that the foregoing procedure is simply a matter of obtaining the 'constants' in (12) by computation; graphical methods do not enter into it, any more than they do in obtaining the arithmetic mean of a number of observed values of a single quantity.

Computation is usually a more accurate process than graphical methods, even in the case of a linear equation which is most favourable to the latter; but it is not so illuminating, or perhaps so convincing. So it may be helpful to show graphically the final result of the above computation.

The two straight lines represented by equations (12) and (13) are shown in Fig. 4*b*; since the vertical and horizontal scales are not equal this is a distorted graph (see §3), but if equal scales had been used the two lines would have been at right angles to one another.

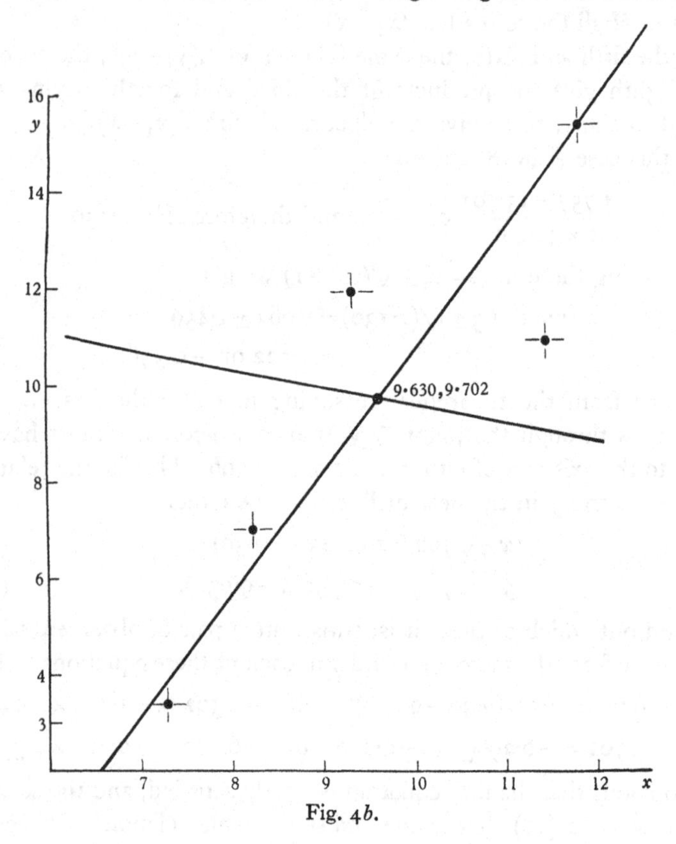

Fig. 4*b*.

It is obvious that the preparation of this table and the deduction of (12) from it is more laborious than plotting the observations, as in Fig. 4*b*, and drawing the best-fit straight line among them 'by eye'. But the result is far more accurate, even if the observed points lie nearly on a straight line; it is foolish to take elaborate precautions for accuracy in the setting up and conduct of the experiment and then to 'spoil the ship for a haporth of tar' by neglecting to use the most accurate method of getting the final result from the observations.

PART I. MECHANICS

Exp. 1. *Stretched elastic string, I*

We know from experience that if we fix a string, rope, wire, etc., to two points at the same level, and stretch it as tightly as we like, and then hang a weight on its centre, it will sag to a certain extent. The purpose of this and the next experiment is to find out how the amount of this sagging depends on the length of the string, its extensibility, the tension we put in it before hanging on the weight, and the magnitude of that weight.

(*a*) Attach a light scale-pan, whose weight (which should be less than 7 g.) you have measured, to a spiral steel spring, which should be not less than 7 cm. or more than 20 cm. long and should give an extension between 0·5 and 1·5 cm. per 100 g. wt. Hang the spring vertically and plot a graph connecting the length (l cm.) of the spring and the load (T g. wt.) on it, including the weight of the scale-pan, up to about 60% extension. l must be measured accurately between two well-defined points on the spring.

(*b*) Drive stout nails, up to half their length, into the edge at A and B of a board about a metre long, and fix the board with its face horizontal. Attach the spring to A and B by strong thread (fine wire, such as S.W.G. 26, will serve) which is strong enough to carry 1 kg., stretching the spring to about $1\frac{1}{2}$ times its unstretched length. Guard against any possibility of the attachments to the nails slipping, by taking several tight turns round the nail before fastening off the thread. Measure the length ($2L$ cm.) of AB, and the stretched length (l_1 cm.) of the spring; deduce from your graph the tension (T_1 g. wt.) of the string and thread corresponding to this length l_1 cm.

Hang the scale-pan, by a loop of cotton through which the thread passes, from a point B on the thread vertically below C, the mid-point of AB, and fix a vertical millimetre scale with its edge passing through C. Measure the vertical displacements (y cm.) of P for a series of loads (w g. wt.). The loop of cotton must be moved along the thread as required to keep P vertically below C. y should not exceed 4 or 5 cm.

Plot a graph with w as ordinates and y as abscissae.

Theory (see appended note) shows that for very small values of y/L this graph should be a straight line through the origin; determine from your graph the slope to the axis of y of this line, if it is a straight line, or if the graph is curved the slope of its tangent at the origin. This 'slope' must be measured, not by a protractor, but by using the scales along the axes; see Introduction (§ 3).

Compare your value of this slope with its theoretical value $2T_1/L$.

(c) Drive another nail into another part of the board's edge, about $\frac{1}{3}L$ from B, and repeat part (b) of the experiment.

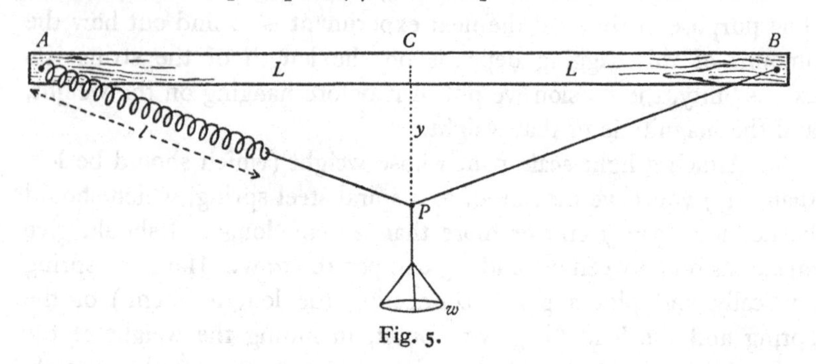

Fig. 5.

If in each case your results agree (except for differences that can legitimately be ascribed to casual errors of experiment) with the theoretical relationship $w=\frac{2T_1}{L}y$, they may be taken to establish that relationship, since the values of T_1 and L were taken at random.

In that case you have proved experimentally that the force needed to displace transversely, through a certain small distance, the centre of a stretched elastic string is directly proportional to its tension before displacement, and inversely proportional to its length, and that if all other things remain unchanged the force is directly proportional to the displacement.

Note on the Theory

Suppose that the spiral spring used in the experiment, when the tension T g. wt. exceeds a certain amount, obeys the law $l=aT+b$, so that an additional tension of 1 g. wt. causes an additional extension of a cm. If the initial tension of the thread is T_1 g. wt., and its tension when P is displaced y cm. is T g. wt., the spring will have increased in length by $a(T-T_1)$ cm. and the length of PA or PB will be $L+\frac{1}{2}a(T-T_1)$ cm.

Denote the angles PAB and PBA in Fig. 5 by β; then

$$y=\left\{L+\frac{a}{2}(T-T_1)\right\}\sin\beta$$

$$=\left(L-\frac{aT_1}{2}\right)\sin\beta+\frac{a}{2}T\sin\beta.$$

Now since P is in equilibrium

$$w=2T\sin\beta.$$

Hence $$y=\left(L-\frac{aT_1}{2}\right)\sin\beta+\frac{aw}{4},$$

or $$\sin\beta=\frac{4y-aw}{2(2L-aT_1)}.$$

But $\tan\beta=\frac{y}{L}$, so that $\sin\beta=\frac{y}{\sqrt{(L^2+y^2)}}$. Expanding this by the binomial theorem

$$\sin\beta=\frac{y}{L}\left(1+\frac{y^2}{L^2}\right)^{-\frac{1}{2}}=\frac{y}{L}\left(1-\frac{y^2}{2L^2}+\frac{y^4}{8L^4}-\right), \text{ etc.}$$

Hence $$\frac{4y-aw}{2(2L-aT_1)}=\frac{y}{L}-\frac{y^3}{2L^3} \text{ nearly},$$

or $$4y-aw=2(2L-aT_1)\frac{y}{L}-(2L-aT_1)\frac{y^3}{L^3} \text{ nearly},$$

or $$w=\frac{2T_1}{L}y+\frac{2L-aT_1}{aL^3}y^3 \text{ nearly}$$

$$=\frac{2T_1}{L}y \text{ if } \frac{y}{L} \text{ is very small.}$$

It will be seen that there is no need here to take into account the difference between the real and the ideal unstretched length of the spring, dealt with in the Introduction (§ 2 (a)), since the constant in the final equation involves only L and the initial tension T_1 which is read off the first graph.

Exp. 2. *Stretched elastic string, II*

Take the set-up of Exp. 1, but make $2L$ about 60 cm., use a spring which stretches about 0·5 or 0·8 cm. per 100 g. wt., and adjust the initial tension (T_1 g. wt.) so that it is not much greater than is required to separate all the consecutive turns of the spiral spring.

Measure $2L$, determine T_1 as in Exp. 1 and the extension (a cm.) per g. wt. for loads more than sufficient to separate consecutive turns, by hanging up the spring as in part (*a*) of Exp. 1, loading it with a succession of weights w, drawing a graph, and measuring its slope to the axis of w, which gives a.

Get the smooth graph connecting w with y as in part (*b*) of Exp. 1, carrying the value of y from 0 to 7 or 8 cm.

Draw tangents to this smoothed graph at points corresponding to values of y of 6, 5, 4, 3 and 2 cm. Measure the slope to the axis of y of the tangent in each case (not by a protractor but by using the scales along the axes; see Introduction (§ 3)), and plot a graph connecting these values as ordinates with the corresponding values of y^2 (*not* of y) as abscissae.

The appended note on theory shows that the graph should be a straight line, making an intercept on the vertical axis of $2T_1/L$, and sloped to the axis of y^2 at $3 \times \dfrac{2L - aT_1}{aL^3}$.

Calculate the intercept and slope of your second experimental graph, and compare them with these theoretical values, after substituting the measured values of L, T_1 and a.

Note on the Theory

In the note appended to Exp. 1 the theoretical relation between w and y was shown to be $w = \dfrac{2T_1}{L}y + \dfrac{2L - aT_1}{aL^3}y^3$ for small values of y/L.

If we keep T_1, L and a unchanged, and differentiate both sides of this equation with respect to y, we get for the slope (dw/dy) to the axis of y of the curve, at the point corresponding to y,

$$\frac{dw}{dy} = \frac{2T_1}{L} + 3 \times \frac{2L - aT_1}{aL^3}y^2.$$

Hence if we measure this slope at a number of points on the graph, and plot it as abscissa against the corresponding value of y^2 as ordinate, we should get a straight line, which should make an intercept $2T_1/L$ on the axis of dw/dy, and should be inclined to the axis of y^2 at a slope of $3 \times \dfrac{2L - aT_1}{aL^3}$.

This is a device for getting a linear graph which is not mentioned in the Introduction.

Exp. 3. *Rigid pendulum*

The 'ideal simple pendulum' is found only in text-books; the simple pendulum found in laboratories is chiefly of use as a cheap and convenient time-measurer. The pendulum of a clock, or any rigid object that can swing in a vertical plane about a fixed horizontal axis under the action of gravity, can be of almost any shape, and the purpose of this experiment is to find out something of the way in which its time of swing depends on that shape and on the distance between its centre of gravity and its point of support.

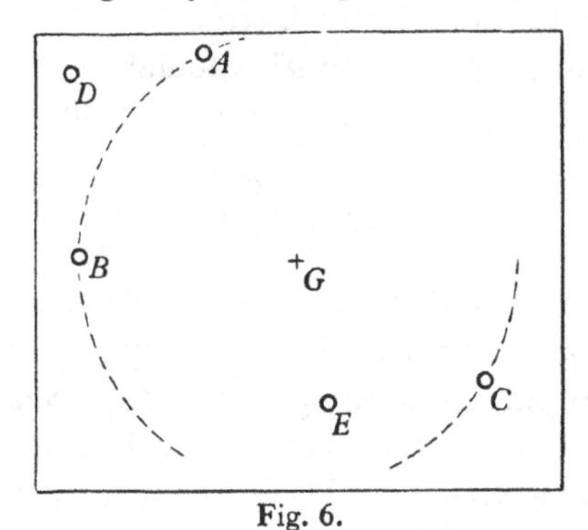

Fig. 6.

Take as the object to be used as the rigid pendulum a piece of cardboard of any shape and size, the larger the better, but a rectangle with sides of 20 and 24 cm. is convenient. Locate its centre of gravity, G, by balancing it on the point of a needle. Punch holes, about 3–5 mm. in diameter, in it as in Fig. 6, the centres A, B and C of three of them being equidistant from G, and D and E wherever you like. Stick the needle through a small cork and support it horizontally by fixing the cork in a retort stand. Pass the needle through A, so that the cardboard can oscillate freely in its own plane like a pendulum; measure the distance (l_1 cm.) of G from the point of support, estimating tenths of a millimetre.

Set up in front of the cardboard a simple pendulum, with its thread held between two pieces of wood in the jaws of the same or another retort stand, so that its length (l_2 cm.) from the centre of gravity of its bob to its point of support can be easily varied and accurately measured. Adjust l_2 until the two pendulums, when displaced from their equilibrium positions and released simultaneously, have the same periods of oscillation, so that they continue to swing together until the oscillations die out.

Measure l_2, the length of the so-called 'Simple Equivalent Pendulum'. Keeping l_2 unchanged, use B and C in succession as points of support of the cardboard, and see whether the cardboard still oscillates in time with the simple pendulum of length l_2 cm.

If so, calculate the value of a^2, where

$$l_1(l_2-l_1)=a^2. \tag{1}$$

Find in the same way the lengths of the simple equivalent pendulums when the holes D and E are used; calculate a^2 in each case, changing l_1 and l_2 in (1) to suit each case.

If, as is probable, the value of a^2 is found to be about constant, calculate the arithmetic mean of its observed values. Then from (1)

$$l_2=l_1+\frac{a^2}{l_1},$$

or, in words, the length of the simple equivalent pendulum exceeds the distance (l_1) between the centre of gravity and the point of support of the rigid pendulum by a^2/l_1, where a^2 is a constant.

If the cardboard is a uniform rectangle and the effect of the holes is ignored, this constant a^2 should theoretically equal one-twelfth of the sum of the squares of the length and breadth of the cardboard. Check this in your case.

Note on the Theory

Text-books of Mechanics show that the period of oscillation of a rigid pendulum equals $2\pi\sqrt{\dfrac{m(l_1^2+k^2)}{ml_1g}}$, where m is its mass, l_1 the distance of its centre of gravity from its point of support, and mk^2 is its moment of inertia about an axis through its centre of gravity perpendicular to its plane of oscillation. So a simple pendulum of length l_2 has a period of oscillation of

$$2\pi\sqrt{\frac{l_2^2}{l_2g}} \quad \text{or} \quad 2\pi\sqrt{\frac{l_2}{g}}.$$

Hence in this case

$$\frac{l_1^2+k^2}{l_1g}=\frac{l_2}{g} \quad \text{or} \quad k^2=l_1l_2-l_1^2=l_1(l_2-l_1) \quad \text{or} \quad l_2=l_1+\frac{k^2}{l_1}.$$

These text-books also show that in the case of a rectangle of sides b and c, $k^2=\frac{1}{12}(b^2+c^2)$.

Exp. 4. *Simple pendulum*

The object of this experiment is to investigate the relation between the length (l) and the period (T) of a simple pendulum, with the help of a graduated ruler, but without using a stop-watch or any theory of mechanics in general or S.H.M. in particular.

Set up a simple pendulum of any length (l_0), but it is convenient to take it about 20 cm.; measure l_0. This pendulum will serve as a standard, and its period as a time-unit, for the experiment.

Set up another simple pendulum, of length between 78 and 82 cm., with its bob at about the same level as the bob of the first and an inch or two in front of it; the line joining the two bobs, when they are hanging at rest, should be roughly perpendicular to the planes in which they will swing. This pendulum can best be supported by passing its thread between two thin pieces of wood gripped between the jaws of a retort stand, so that the length (l) of the pendulum can be adjusted by sliding the thread between the pieces of wood when the jaws are slightly opened.

Start the two pendulums simultaneously by 'kicking them off' gently with the face of a ruler held against them while they hang freely at rest. Adjust the length of the front pendulum so that the two pendulums pass simultaneously through their initial central positions after one complete, to-and-fro, swing of the front pendulum and two complete swings of the back one.

If you have made this adjustment with absolute precision, the two pendulums will coincide after every whole number of swings of the longer pendulum, and you should improve your adjustment accordingly; in fact, you can in this way compare the periods of the two pendulums more accurately than you can compare their lengths. It corresponds to timing a large number of swings when using a stop-watch. Measure l, the length of this front pendulum.

Repeat the experiment, adjusting the length of the front pendulum so that it makes two complete swings while the standard one makes three swings; expressing it otherwise, it has to make n swings while the standard one makes $n+1$ swings, n being two in this case, and one in the first case.

Repeat the experiment with n equal to, say, 3 and 4.

Make a table of your corresponding values of n and l. These quantities must be connected by some law; the next step is to discover it. This can best be done by assuming various forms which the law is likely to take, and testing each by applying it to the four pairs of values of n and l which you have obtained.

The simplest hypothesis is that $l=An^p$, where A and p are constants. If this is so,

$$\log l = \log A + p \log n,$$

and the graph connecting $\log l$ and $\log n$ should be a straight line. Test this by plotting the graph connecting your values of $\log l$ and $\log n$; you will probably get a curved graph, and if so you must reject this hypothesis and try out another.

We have found that one pendulum, of variable length l, makes n swings in the same time as another, of fixed length l_0, makes $n+1$ swings; a possible form of the law would seem to be that

$$ln^p = l_0(n+1)^p \quad \text{or} \quad \frac{l}{l_0} = \left(\frac{n+1}{n}\right)^p$$

or

$$\log \frac{l}{l_0} = p \log \left(\frac{n+1}{n}\right),$$

where p is some fixed number.

In that case, the graph connecting $\log l/l_0$ and $\log\left(\frac{n+1}{n}\right)$ should be a straight line through the origin, with a slope of p to the axis of $\log\left(\frac{n+1}{n}\right)$.

You can test this by your experimental results; add to your table columns showing $\log l - \log l_0$ and $\log(n+1) - \log n$, and plot the graph connecting the last two columns. You will probably find that this hypothesis is confirmed, and that the most probable value of p is 2·0.

If so, the relation is

$$\log \frac{l}{l_0} = 2 \log \left(\frac{n+1}{n}\right)$$

or

$$\frac{l}{l_0} = \left(\frac{n+1}{n}\right)^2. \tag{1}$$

Since you took n more or less at random and (1) holds good in all the cases you tried, it is safe to assume that it will hold good for any value of n which is a whole number.

But we want the relation between the length and period of any pendulum, not the relation between l and n under these conditions; so a further step is required.

Denote by T sec. the period of any of the pendulums for which (1) holds good, and by T_0 sec. the period of the standard (back) pendulum. Since the front pendulum makes n swings while the standard pendulum makes $n+1$ swings, nT must equal $(n+1)T_0$, or

$$\frac{n+1}{n}=\frac{T}{T_0}, \tag{2}$$

where n is any whole number. Substituting this in (1) we get

$$\frac{l}{l_0}=\left(\frac{T}{T_0}\right)^2$$

or

$$\frac{T^2}{l}=\frac{T_0^2}{l_0}=A, \tag{3}$$

where A is some constant which does not depend on T or l.

In this form, (3) is a general law connecting T and l in any simple pendulum. But it is very important to realise that you have not really 'verified' this law by your experiments; all that you have done is to discover and verify it in a limited number of cases. Even in those cases you did not actually measure T, but arrived at its value, quite legitimately, by comparison with T_0.

Strictly speaking there is the same imperfection in the ordinary laboratory method of verifying the law by measuring the periods of simple pendulums of various lengths by means of a stop-watch. For the second hand of a watch does not move uniformly, but by a series of jerks; it will only measure an interval of time that begins and ends with these jerks. It is most unlikely, if l is taken at random, that the corresponding T will be exactly an interval of this kind. For a theoretically sound verification we must use a time-measurer such as a chronograph which moves at uniform speed. In practice, the error caused by using a stop-watch is so small that it is less serious than 'casual' errors of observation, and the theoretical imperfection of this practical method of verifying the law is overlooked.

But it is too large in this case to be overlooked. The difficulty cannot be overcome by the argument, which in many cases is valid, that you took l_0 at random, so the law may be expected to hold good for any value of l_0 and therefore of l. For it is quite possible that the constant A in (3) may depend on the mass or length or period of the standard pendulum, and thus may not be the same for different groups of front pendulums. However, the appended note on theory shows that this is not the case, if we can trust the theoretical reasoning, and this argument applies also to the verification of the law by means of a stop-watch. Hence we can claim in both cases alike that we can verify the law by a combination of experiment and theoretical reasoning.

Note on the Theory

We have shown experimentally that for any simple pendulum in a certain group, of length l and period T, $T^2=Al$, where A does not depend on T or l; the only other physical quantities on which A for any group can depend are the acceleration (g) of gravity at the place of experiment and the period (T_0) and the mass (m_0) of the bob of the standard pendulum of its particular group. We can therefore represent A by $Nm_0^q g^r T_0^s$, where N, q, r and s are numbers.

Then $$T^2 = Nm_0^q g^r T_0^s l \qquad \text{(i)}$$

is an equation which holds good for every simple pendulum.

Now it is theoretically necessary that in such an equation every term should have the same dimensions in mass, length and time, separately. The dimension of the left-hand term of (i) in mass is zero, and of the right-hand term is q; hence $\mathbf{q=0}$.

The dimension of the left-hand term in length is zero, and of the right-hand term is $r+1$; hence $r+1=0$, or $\mathbf{r=-1}$.

The dimension of the left-hand term in time is 2, and of the right-hand side is $-2r+s$; hence $-2r+s=2$, or since $r=-1$, $\mathbf{s=0}$.

Consequently, $A=N/g$ and does not depend on T_0 or m_0; so it is the same for every group, and every simple pendulum obeys the same law

$$T^2 = \frac{Nl}{g},$$

where N is some number.

Exp. 5. *Bending of a lath, I*

The object of this and the two following experiments is to check the theoretical reasoning (which is far from simple) by which we can determine the behaviour of a straight elastic beam, supported horizontally at two points and loaded at a third point with sufficient weight to bend it to a moderate extent. If you did not have this theory as a guide, any experiments you might make would probably not be very illuminating; but using it as a guide your experiments should establish the truth of a rather complicated and useful formula.

The following apparatus will be needed in this and the next two experiments; two short glass tubes or rods supported horizontally and parallel to one another on piles of wood blocks at equal heights of about 1 ft. above the bench; the tubes should be prevented from moving on the blocks by means of tin-tacks, and the blocks must give them firm support which does not yield appreciably when loaded. A boxwood metre ruler, lying face upwards on the tubes; the ruler must not rock on the tubes. A scale-pan, which may be made of a piece of plywood about 6 in. square, which can be hung by a thread passed over the ruler at any desired point *P*; a set of weights up to a total of 1000 or 1500 g.; a pin which can be fixed

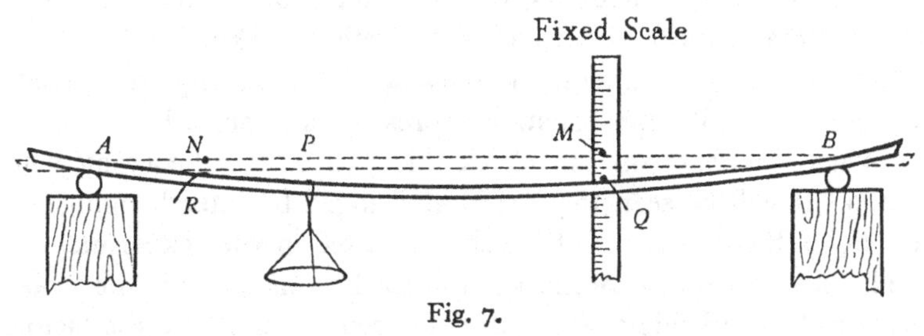

Fig. 7.

at a point *Q* on the upper face of the ruler by a piece of soft wax so that it overhangs the edge of the ruler, and a scale held vertically in a retort stand, with which to measure the vertical displacement (y cm.) of *Q* below its position (*M*) when the scale-pan carries no load (tenths of a millimetre to be estimated).

Fix *A* and *B* about 94 cm. apart; hang the scale-pan from a point *P* on the ruler, about 35 cm. from *A*, and keep *P* unchanged throughout the experiment. Fix a pin at a point *Q*, about 55 cm. from *A*, keeping *Q* unchanged throughout the experiment. Load the scale-pan with an increasing, and then decreasing, weight W g., disregarding the weight of the scale-pan, and observe the corresponding values of y up to a maximum of 2 or 3 cm. Plot a graph connecting y and W. It will probably be a straight line through the origin.

Repeat the operation, hanging the scale-pan from *Q* and measuring vertical displacements of *P*. Plot these observations of y and W on the same axes; the corresponding graph will probably be almost identical with the former graph. If so, *P* and *Q* are interchangeable; the 'deflection' of the ruler at any point *P* caused by a given load on the ruler at any other point *Q* is the same as the deflection at *Q* caused by the same load at *P*.

Hence, if we want to find the deflection at every point of the ruler caused by a given load at a fixed point P we can do so by setting up the pin and scale at P and moving the given load from point to point along the ruler, observing the deflection at P in each case. In this way we could, without moving the vertical scale, find experimentally the shape of the whole length of an initially straight ruler when loaded with a constant weight at P.

The appended note shows that theoretically

$$y = \frac{2Wg}{Ecd^3}\frac{AP \times BQ}{AB}(AB^2 - AP^2 - BQ^2), \qquad (1)$$

where E is Young's modulus for the material of the ruler, c cm. is its horizontal breadth and d cm. is its vertical thickness.

If, as in this experiment, the only factors in the equation which vary are y and W, the equation represents a straight line through the origin, so that you can check this part of the theory by your results. It will be seen that (1) is unchanged by interchanging AP and BQ, a theoretical result which you tested in your experiment.

The slope of the theoretical graph to the axis of W is the whole of the right-hand side of (1) with the exception of W; all the factors of it, except E, are measurable (d should be measured with a screw-gauge); hence, if you measure the slope of your linear graph to the axis of W, you can calculate the value of E. If you used a boxwood ruler you will probably get a value of about $1 \cdot 3 \times 10^{11}$.

Note on the Theory

The mathematics required for getting the equations to the curves formed by an initially straight beam when loaded at one point is too long to be set out here in full, but it is given in Appendix C for the use of those who are sufficiently interested to study this piece of theory. It is there shown that the equation of the curve PQB in Fig. 7, if Young's modulus for the material of the beam is E and if the cross-section of the beam is a rectangle, of horizontal breadth c cm. and vertical thickness d cm., and if the beam is loaded at the point P with W g. wt., is

$$y = \frac{2Wg\,AP\,x}{Ecd^3\,AB}(AB^2 - AP^2 - x^2), \qquad (1)$$

where $x = MB$, $y = QM$ are the co-ordinates of the point Q on the curve PQB.

Similarly, the equation of the curve PRA is

$$\eta = \frac{2Wg\,BP\,\zeta}{Ecd^3\,AB}(AB^2 - BP^2 - \zeta^2), \tag{2}$$

where $\zeta = NA$, $\eta = RN$ are the co-ordinates of the point R on the curve PRA, N being the foot of the perpendicular from R on AB.

The main object of this and the two following experiments is to check this theory; or, if the value of E is not known, so that to this extent the theory cannot be checked, to assume the truth of the theory and use it to determine E.

Exp. 6. *Bending of a lath, II*

Using the same apparatus as described for Exp. 5, fix the pin at P and set up the scale for measuring its displacement; AP is to be kept unchanged throughout the experiment, and it is convenient to make its length about two-fifths of the length of the ruler. The ruler should not overhang A by more than 1 or 2 cm.

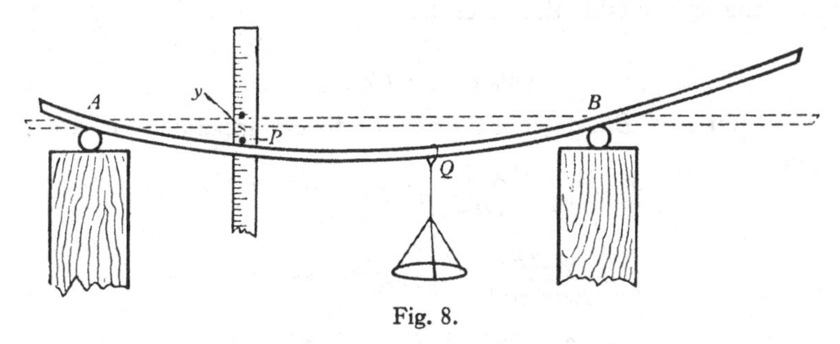

Fig. 8.

Adjust B so that AB is about two-thirds of the length of the ruler. Hang the scale-pan from a point Q between P and B, BQ being $1/n$th of AB; n is to be kept constant throughout the experiment, and it can conveniently be between 3 and 4; it need not be a whole number.

Observe y, the displacement of P caused by putting some weight W in the scale-pan; W should be chosen so that y lies between 1 and 2 cm., to get an easily readable value without unduly straining the ruler.

Measure AB and record the values of y/W and AB^2.

Readjust the position of B to get a larger value of AB; shift Q, the point of support of the scale-pan so that BQ again becomes AB/n; observe y for the same value of W. If y becomes unduly large when

AB is increased, W may be reduced, since by Exp. 5 the value of y/W does not vary with W. Record the new values of y/W and AB^2.

Repeat this for two or three other positions of B.

Plot the graph connecting y/W and AB^2.

The appended note on theory shows that the relation between y/W and AB^2 should theoretically be

$$\frac{y}{W}=\frac{2g\,AP(n^2-1)}{n^3\,Ecd^3}\left(AB^2-\frac{n^2}{n^2-1}AP^2\right).$$

Since AP, n, E, c and d are constants the graph representing this relation should be a straight line, and this can be checked from the observed graph. The intercept on the axis of AB^2 should theoretically be $\frac{n^2}{n^2-1}AP^2$, and this value can be checked from the graph and the known values of n and AP.

So this experiment enables us to check completely the part of the complete theoretical equation inside the brackets, and to some extent the part outside the brackets.

Note on the Theory

If we put $x=AB/n$ in (i) of the note on theory appended to Exp. 5, we get

$$\frac{y}{W}=\frac{2g\,AP\times AB}{Ecd^3\,AB\,n}\left(AB^2-AP^2-\frac{AB^2}{n^2}\right)$$

$$=\frac{2g\,AP}{Ecd^3\,n^3}\{(n^2-1)\,AB^2-n^2AP^2\}$$

$$=\frac{2g\,AP(n^2-1)}{n^3\,Ecd^3}\left(AB^2-\frac{n^2}{n^2-1}AP^2\right).$$

Exp. 7. *Bending of a lath, III*

Using the same apparatus as described for Exp. 5, adjust it so that AB, AP, BQ and BD are respectively about two-thirds, one-third, one-quarter and one-quarter of the length of the ruler. Fix pins at Q and D, and set up scales to measure their vertical displacements, y and z respectively; hang the scale-pan at P.

Put a weight in the scale-pan (its magnitude need not be known), and take the readings of Q and D before this load is added, while it is in the pan, and (to check that there is no permanent distortion of the ruler) after the weight is removed. Note the values of y and z caused by this load.

Repeat with different loads, keeping everything else unchanged. Plot a graph connecting y and z. If, as is probable, it is a straight line for moderate loads, calculate its slope to the axis of z.

The appended note on theory shows that this graph should theoretically be a straight line with a slope to the axis of z, whose value is

$$\frac{BQ}{BD}\left(1-\frac{BQ^2}{AB^2-AP^2}\right).$$

Measure these lengths, calculate this theoretical slope and compare its value with the slope of your graph.

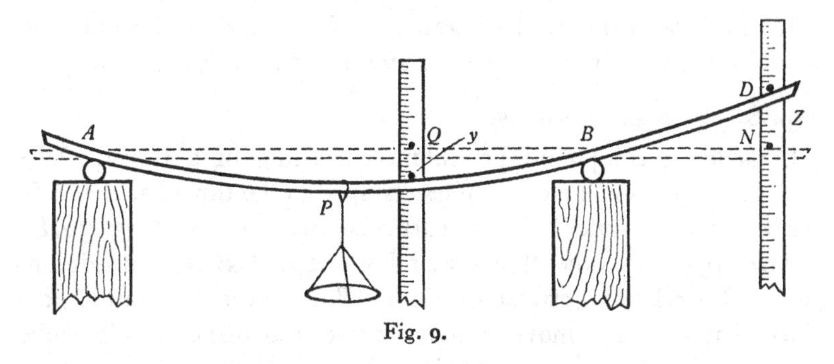

Fig. 9.

Hence, by combining the results of Exps. 5, 6 and 7, you can check experimentally the whole of equation (1) in the note appended to Exp. 5, with the exception of the value of Young's modulus for the material.

Note on the Theory

There is no bending moment at any point of the overhanging part BD of the ruler, so this part remains straight and has the same inclination to the horizontal as the tangent of the curve at B; since this inclination is small it is nearly equal (in radians) to DN/BD.

The tangent of the angle of inclination of the tangent at any point x, y of the curve (1) of the note to Exp. 5 is got by differentiating the right-hand side of (1) with respect to x, so it is

$$\frac{2Wg\,AP}{Ecd^3\,AB}(AB^2-AP^2-3x^2).$$

Its value at $x=0$ is therefore

$$\frac{2Wg\,AP}{Ecd^3\,AB}(AB^2-AP^2).$$

Hence DN, or z, equals

$$\frac{BD}{Ecd^3}\frac{2Wg\,AP}{AB}(AB^2-AP^2). \tag{1}$$

But from (1) of that note, when $x=BQ$,

$$y=\frac{2Wg\,AP\times BQ}{Ecd^3\,AB}(AB^2-AP^2-BQ^2). \tag{2}$$

Hence, dividing (2) by (1),

$$\frac{y}{z}=\frac{BQ}{BD}\frac{AB^2-AP^2-BQ^2}{AB^2-AP^2}=\frac{BQ}{BD}\left(1-\frac{BQ^2}{AB^2-AP^2}\right).$$

It will be seen that we have got rid of E, c and d, so this equation is applicable to a uniform beam of any material or dimensions.

Exp. 8. *Compound pendulum*

Consider a compound pendulum consisting of a thread OAB fixed at O, with small equal masses clamped to the thread at A and B. We will here consider only the particular case when $OA=3{\cdot}25\,AB$.

The appended note shows that both A and B in a pendulum having this relation between OA and AB, however it is started provided that A and B move in the same vertical plane, should move with a combination of two S.H.M.'s, and that if T_1 is the periodic time of the fundamental S.H.M. and T_2 that of the upper harmonic or overtone, then

$$T_1^2=\frac{4\pi^2\,3{\cdot}82}{g}AB \quad \text{and} \quad T_2^2=\frac{4\pi^2\,0{\cdot}425}{g}AB.$$

It will be seen from the above that $T_1=3T_2$.

The amplitudes of the component S.H.M.'s depend on the positions of A and B when the oscillation starts. In particular, if OAB is then nearly a straight line, that is, if B is held before it is released so that the thread OAB is nearly but not quite taut as in Fig. 10, the amplitude of the overtone will be so small that we can disregard it, and both A and B will oscillate with the periodic time T_1 of the fundamental.

If, however, A and B are held as in Fig. 11 on opposite sides of the vertical line OV in such a position that the point P where AB cuts OV is at a distance of $0{\cdot}425\,AB$ from B, and if A and B are then released simultaneously, the fundamental will be suppressed and both A and B will oscillate with the periodic time T_2 of the overtone.

We can test the truth of these theoretical results as follows.

The masses can most conveniently be made by cutting two rectangles, less than 4 by 8 mm., out of sheet lead about 1·5 mm. thick. They should be made of very nearly equal weight by trimming down the heavier with a knife. They are then to be folded in half and pinched on the thread at the desired points by a pair of pliers or a hammer; the thread must pass nearly through the centre of each square. We can conveniently make AB about 25 cm. in the first place, and OA exactly 3·25 AB.

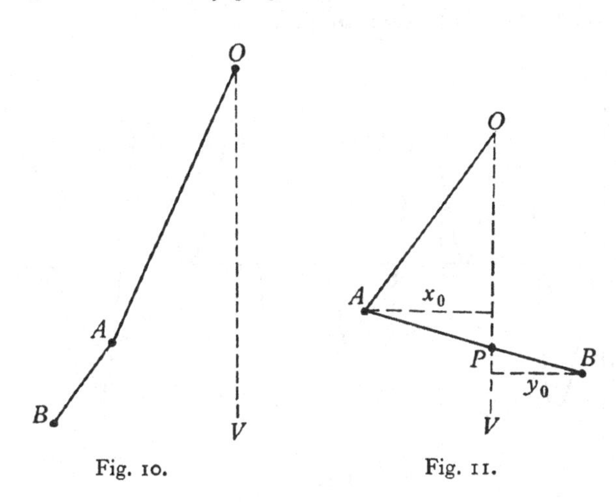

Fig. 10. Fig. 11.

To determine T_1 in terms of AB. Start the pendulum by holding B about 6 in. from OV, with OAB not quite taut; release and measure with a stop-watch the periodic time T_1 of A or B; this will be the fundamental period. Record T_1 and the corresponding value of AB.

Open the leaves of A with a knife and clamp it again at a different point of the thread; adjust O to give $OA = 3{\cdot}25\ AB$. Measure T_1 with this value of AB.

Repeat for two or three other values of AB, and plot a graph of T_1^2 against AB; if it is a straight line compare its slope to the axis of AB with the theoretical value $\frac{4\pi^2\, 3{\cdot}82}{g}$.

You can thus test both the theoretical relation $T_1^2 = K \times AB$, and the theoretical value of K in this particular case of $OA = 3{\cdot}25\ AB$.

To determine the relation between T_1 and T_2. Using any pendulum for which you have measured AB and T_1, stick a tiny bit of gummed paper, in the same way as you fixed the lead to the thread, at a point P of AB such that $PB = 0{\cdot}425\ AB$; this will act as a mark.

Hold A and B on opposite sides of OV, as described above, so that P in Fig. 11 is vertically below O and OA and AB are taut, and release them simultaneously. If you have done all this with complete precision, P should remain in OV throughout the motion. Measure with a stop-watch the periodic time T_2 of A or B, and test whether $T_2 = \frac{1}{3}T_1$ for this pendulum.

Since in these experiments you have taken AB at random, if they verify the theory for some particular value of AB they can be taken to verify it for all values of AB.

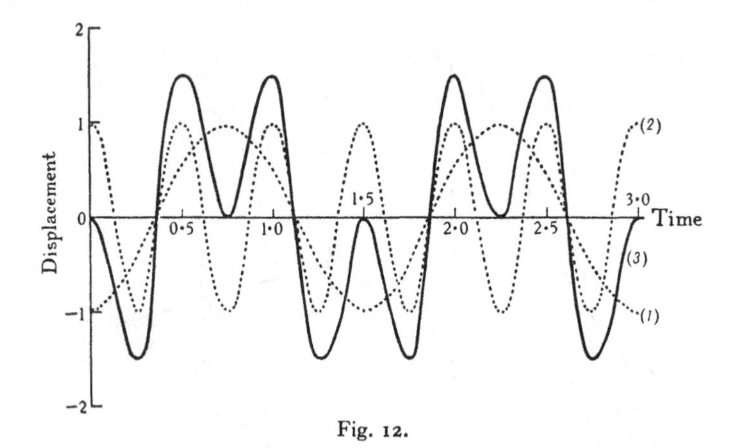

Fig. 12.

Composition of two S.H.M.*'s*. Using any of your compound pendulums (the longer the better), hold A in OV and B about 6 in. from OV, with OA and AB taut; release A and B simultaneously.

Theoretically, the distance of A from OV throughout the subsequent motion should be represented by the ordinate of the full-line graph (3) in Fig. 12, in which abscissae represent times. In this graph the ordinates are the algebraic sums of the ordinates at the same instant of the dotted S.H.M.'s marked (1) and (2). Compare this with the motion in your experiment.

Can you, after studying Fig. 12, measure T_1 in this case? If so, compare it with T_1 for the same pendulum, started as in the first case.

Note on the Theory

Suppose that two equal masses, m, whose moments of inertia are negligibly small, are attached at A and B to a fine thread OAB, fixed at O, and that OA and AB make very small angles, θ and ψ, with the vertical OV. Denote AB by l and OA by al, and the tension in OA by T and the tension in AB by T'.

Then resolving vertically

$$T' = mg, \quad T - T' = mg,$$

and horizontally

$$mal\frac{d^2\theta}{dt^2} = T'\psi - T\theta,$$

$$m\left(al\frac{d^2\theta}{dt^2} + l\frac{d^2\psi}{dt^2}\right) = -T'\psi.$$

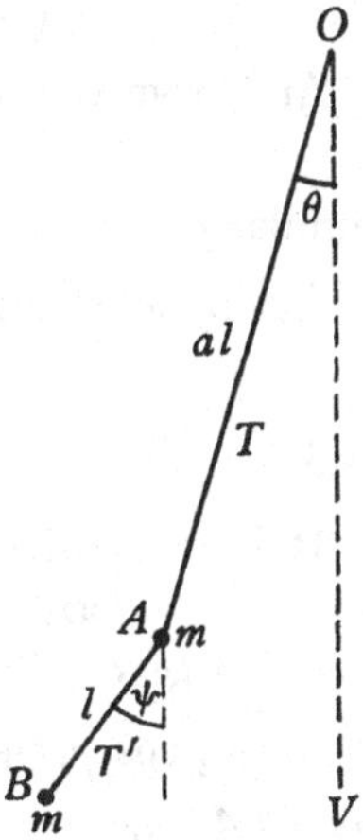

Eliminating T and T' and writing D^2 for d^2/dt^2 we get

$$(alD^2 + 2g)\,\theta - g\psi = 0, \qquad (1)$$

$$alD^2\theta + (lD^2 + g)\,\psi = 0. \qquad (2)$$

Fig. 13.

Therefore

$$\{(alD^2 + 2g)(lD^2 + g) + algD^2\}\ \theta \text{ or } \psi = 0$$

or

$$\{al^2D^4 + 2gl(1+a)D^2 + 2g^2\}\ \theta \text{ or } \psi = 0. \qquad (3)$$

Hence

$$-D^2 = \frac{g}{al}\{1 + a \pm \sqrt{(1 + a^2)}\}.$$

Denote

$$\frac{g}{al}\{1 + a + \sqrt{(1 + a^2)}\} \text{ by } \gamma_1^2 \quad \text{and} \quad \frac{g}{al}\{1 + a - \sqrt{(1 + a^2)}\} \text{ by } \gamma_2^2. \qquad (4)$$

Then $\theta = A_1 \cos \gamma_1 t + B_1 \sin \gamma_1 t + A_2 \cos \gamma_2 t + B_2 \sin \gamma_2 t$ (5)

and $\psi = C_1 \cos \gamma_1 t + D_1 \sin \gamma_1 t + C_2 \cos \gamma_2 t + D_2 \sin \gamma_2 t$ (6)

satisfy the equation (3) whatever values the eight constants, A, B, C and D, may have.

These values of θ and ψ must satisfy (1) for all values of t, so that

$$(-al\gamma_1^2 + 2g)\,A_1 = gC_1, \qquad (7)$$

$$(-al\gamma_1^2 + 2g)\,B_1 = gD_1, \qquad (8)$$

$$(-al\gamma_2^2 + 2g)\,A_2 = gC_2, \qquad (9)$$

$$(-al\gamma_2^2 + 2g)\,B_2 = gD_2. \qquad (10)$$

Suppose that, when $t = 0$, $\theta = \theta_0$, $d\theta/dt = 0$, $\psi = \psi_0$, $d\psi/dt = 0$.

Then from (5)

$$\theta_0 = A_1 + A_2, \qquad (11)$$

$$0 = \gamma_1 B_1 + \gamma_2 B_2, \qquad (12)$$

and from (6)

$$\psi_0 = C_1 + C_2, \qquad (13)$$

$$0 = \gamma_1 D_1 + \gamma_2 D_2. \qquad (14)$$

Hence from (8) and (10) and (14)

$$\gamma_1(-al\gamma_1^2+2g)\,B_1+\gamma_2(-al\gamma_2^2+2g)\,B_2=0.$$

And from this and (12)

$$B_1=B_2=0.$$

Similarly,

$$D_1=D_2=0.$$

Again, substituting for γ_1^2 and γ_2^2 in (7) and (9),

$$\{1-a-\sqrt{(1+a^2)}\}\,A_1=C_1 \tag{15}$$

and

$$\{1-a+\sqrt{(1+a^2)}\}\,A_2=C_2. \tag{16}$$

It is convenient to deal with the displacements, x and y of A and B respectively from the vertical line OV, so that $x=al\theta$ and $y=al\theta+l\psi$, or, from (5) and (6) and (15) and (16),

$$x=al(A_1\cos\gamma_1 t+A_2\cos\gamma_2 t), \tag{17}$$

$$\begin{aligned} y&=al(A_1\cos\gamma_1 t+A_2\cos\gamma_2 t)+l\{[1-a-\sqrt{(1+a^2)}]\,A_1\cos\gamma_1 t\\ &\qquad+[1-a+\sqrt{(1+a^2)}]\,A_2\cos\gamma_2 t\}\\ &=l\{[1-\sqrt{(1+a^2)}]\,A_1\cos\gamma_1 t+[1+\sqrt{(1+a^2)}]\,A_2\cos\gamma_2 t\}. \end{aligned} \tag{18}$$

If x_0 and y_0 are the initial values of x and y,

$$x_0=al(A_1+A_2),$$

$$y_0=l\{[1-\sqrt{(1+a^2)}]\,A_1+[1+\sqrt{(1+a^2)}]\,A_2\}.$$

Hence

$$A_1=\frac{x_0\{1+\sqrt{(1+a^2)}\}-y_0 a}{2al\sqrt{(1+a^2)}} \tag{19}$$

and

$$A_2=-\frac{x_0\{1-\sqrt{(1+a^2)}\}-y_0 a}{2al\sqrt{(1+a^2)}}. \tag{20}$$

If then we substitute these in (17) and (18) we get the general equations to the motion of both A and B in terms of the time and of their initial positions.

Consider the particular case of this experiment, where $\gamma_1=3\gamma_2$, and therefore the whole system has a periodic motion, which repeats itself after γ_1 sec.

Then from (4)

$$1+a+\sqrt{(1+a^2)}=9\{1+a-\sqrt{(1+a^2)}\},$$

and therefore $a=3{\cdot}25$ or $0{\cdot}308$.

Taking the former value of a, so that, as in the experiment, OA is longer than AB; from (4)

$$\gamma_1^2=\frac{2{\cdot}35g}{l} \quad\text{and}\quad \gamma_2^2=\frac{0{\cdot}261g}{l},$$

and from (19)

$$A_1 = \frac{0\cdot198x_0 - 0\cdot147y_0}{l}, \quad A_2 = \frac{0\cdot108x_0 + 0\cdot147y_0}{l}.$$

Hence from (17)

$$x = (0\cdot647x_0 - 0\cdot477y_0)\cos\gamma_1 t + (0\cdot353x_0 + 0\cdot477y_0)\cos\gamma_2 t, \quad (21)$$

and from (18)

$$y = -(0\cdot477x_0 - 0\cdot353y_0)\cos\gamma_1 t + (0\cdot477x_0 + 0\cdot647y_0)\cos\gamma_2 t. \quad (22)$$

If $y_0/x_0 = -0\cdot74$, both $0\cdot353x_0 + 0\cdot477y_0$ and $0\cdot477x_0 + 0\cdot647y_0$ vanish and the fundamental period $2\pi/\gamma_2$ disappears from the oscillations of both A and B; both will then oscillate with S.H.M. of the overtone $2\pi/\gamma_1$.

This happens with the arrangement shown in Fig. 11.

Again, if $y_0/x_0 = 1\cdot35$, both $0\cdot647x_0 - 0\cdot477y_0$ and $0\cdot477x_0 - 0\cdot353y_0$ vanish, and the overtone disappears from the oscillations. This will happen if OAB is almost a straight line initially.

And if $x_0 = 0$,

$$x = -0\cdot477y_0(\cos\gamma_1 t - \cos\gamma_2 t),$$

which is shown in Fig. 12.

Exp. 9. *Resonance*

If a body at rest is acted on by a small periodic force, whose period is exactly the same as the period with which the body would naturally vibrate, the body may gradually attain a state of vibration of large amplitude (perhaps even destructively large, if the periodic force persists for a long time), and will continue to vibrate for some time after the force ceases to act. Thus, if you sing a steady note in front of a piano, and one of the strings is in tune with that note, it will start vibrating and can be heard to go on vibrating after you have stopped singing. Again, a wireless set responds when it is properly 'tuned in'; and so on.

The following experiment is an investigation of a rather less simple case of resonance, where the same body can vibrate in two different ways, with periods which can easily be controlled and measured.

From a really rigid support, such as a nail in the edge of the bench, hang a short and stiff spiral steel spring. A convenient spring would be about 7 cm. long unstretched which doubles its length when loaded with about 900 to 1500 g. Attach a scale-pan by a short length of thread or fine string to the bottom of the spring; a convenient scale-pan can be made of a square of plywood, say 4×4 in., with strings attached near its corners; the overall length from the nail to the bottom of the scale pan should not exceed about 20 cm. Load the pan with 100 g. and set it swinging sideways; the motion will probably remain almost entirely pendular. Increase the load by steps of 100 to 500 g. and note how it oscillates after being started as a pendulum. You will probably observe, for any load, the gradual intrusion of a vertical oscillation, and an exchange from pendular to vertical oscillation and back again, to an extent that depends on the load.

Increase the load until it stretches the spring to about double its unstretched length; if you go beyond this you may produce a permanent set in the spring. With this load you will probably find that vertical oscillations are again hardly perceptible.

Broadly speaking, the overall length of the pendulum, and therefore its period for pendular swings, does not increase much with increased loads, but its period of vertical oscillation is greatly increased; theory shows that the square of this period is approximately proportional to the load. Hence, as you changed the load you changed the ratio between the two periods, and it would seem likely that the exchange between the two modes of oscillation occurs most strongly for some definite numerical ratio between them; in other words, that you have been 'tuning in' one mode of oscillation to the other, hunting for the point at which resonance reaches its maximum.

General theoretical considerations (see attached Note) suggest that resonance should be most marked if the load is such that the pendular period is exactly double the period of vertical oscillation. The load required to produce this effect with this set-up of spring and scale-pan can most easily be determined as follows.

Load the scale-pan to a total of M g. which must be sufficient to separate the turns of the spring even when oscillating vertically. Find the periodic times of vertical and small pendular oscillations (T_1 and T_2 sec. respectively).

Repeat this for two other values of M.

You may assume from theory that, approximately,

$$T_1^2 = A_1 M + B_1 \quad \text{and} \quad T_2^2 = A_2 M + B_2,$$

where A_1, B_1, A_2 and B_2 are constants; hence the graphs of T_1^2 and T_2^2 against M will both be straight lines.

Plot with the same axes and scales a graph connecting $4T_1^2$ and

M, and a graph connecting T_2^2 and M; the value of M corresponding to the intersection of these two graphs will obviously be the load needed to give $2T_2 = T_1$ in the spring-supported pendulum you have set up.

Load the scale-pan to this value of M. Displace the scale-pan horizontally, release it and watch the subsequent motion, in which pendular oscillation will probably gradually change into a vertical oscillation, and then back to a pendular oscillation, and so on.

Increase or decrease the load on the scale-pan by about 50 g., and observe the degree of completeness of the change from one form of motion to the other for this load.

By further experiments of this sort you can find the exact load which gives the most perfect resonance; it may not be the load found from the intersections of the two graphs. You can also investigate the behaviour of the system when the resonance is not complete; for instance, you can determine the 'phase difference' between the component oscillations, which is shown by the position of the pan in its horizontal swing when it is at its lowest point in its vertical oscillation.

Note on the Theory

To determine the motion of the scale-pan under the conditions of this experiment calls for dynamics and mathematics too advanced to be given here. But the experimental results can be explained, up to a point, in words in a general way. The following is an outline of such an explanation; it is far from complete and wide open to destructive criticism, but it does suggest that if there is a certain relation between the periods of vertical and horizontal oscillations the effects observable in the experiment may well happen.

An ordinary pendulum at rest can be made to swing through a large arc by a succession of tiny taps if they are applied in a constant direction as the pendulum is at or near its lowest point; so the interval between taps must equal the natural period of the pendulum. Here the string of the pendulum is inextensible, and if the bob swings freely under gravity in a vertical plane it can only move backwards and forwards in a circular arc, always moving perpendicular to the string; theory shows that the tension of the string changes slightly during this oscillation.

If the string is made elastic, motion along the string becomes possible, and the above changes in the tension of the string will now lead to that form of motion. Hence the foregoing series of tiny horizontal taps which set up a considerable horizontal swing will also lead to a vertical oscillation. This latter motion will remain insignificantly small unless the taps are in unison with the vertical as well as with the horizontal oscillation. Hence, in order that the taps should set up considerable oscillations both vertically and horizontally there must be a definite relation between the two natural periods of these oscillations.

Up to this point we have assumed that the taps are produced by an external agent; they can, however, be regarded as taking place within the system of the pendulum swinging under gravity. For if the bob is initially swinging horizontally there will be changes of tension in the string, which produce a vertical impulse on the bob every time it passes through its lowest point, and this impulse will be in the same direction whether the bob is moving from right to left or from left to right. Hence the interval between successive vertical impulses in the same direction is *one half* the period of horizontal swing of the pendulum. If then the natural period of vertical oscillation of the bob is equal to one half of its natural period of horizontal swing, and if the pendulum is swinging horizontally to begin with, we should expect resonance to occur without any outside agency, and a substantial vertical oscillation to be built up.

The creation of this new oscillation demands a supply of energy, and there is now no external agency from which it can be drawn; it must therefore come from the horizontal swings, with consequent decrease in their amplitude. We may expect then that the horizontal swings will be converted gradually into vertical oscillations, though the process cannot well go on so far as totally to extinguish the horizontal swings. Let us assume that it has gone nearly to this limit; we then have a pendulum performing large vertical and small horizontal oscillations, and we can treat the small horizontal components of the changes of tension of the elastic string as horizontal taps. These will build up considerable horizontal swings if they are suitably timed; we can, in the same way as before, show that the above relation between the natural periods of the two oscillations will give this suitable timing. Hence it would appear that resonance should now occur in the opposite direction, and that the gradual interchange between vertical and horizontal motions which is observable in the experiment should take place.

It is difficult, perhaps impossible, to explain in words why these two transfers of energy go on until they are nearly complete, first one way then the other, as the experiment shows that they do; it would seem on the face of it more likely that they would settle down into some half-way state in which the flows of energy were equal and cancelled each other out, with both forms of oscillation present (on the analogy of certain chemical reactions). But the experiment shows that such is not the case; perhaps the best way to put it in words is to say that under these experimental conditions *any* mode of motion of the system is theoretically 'unstable', like that of a flag which flaps continually in a steady wind.

Exp. 10. *Bifilar suspension*

Fix the centre of a half-metre ruler in a retort stand so that its graduated face lies in a horizontal plane, the heights of its two ends above the bench being made equal. Make two loops of thread, so that the doubled thread is about 30 cm. long, and as nearly equal in the two cases as you can get them when tying the knots. Hang these two loops over the fixed ruler, $2a$ cm. apart, at, say, the 15 and 35 cm. graduations, and hang another half-metre ruler in the bottoms of the loops, so that the four lengths of thread between the rulers are vertical. If the heights of the ends of the movable ruler above the bench differ by more than a few millimetres, you can bring them nearer to equality by inserting a rod or wedge of suitable thickness under the thread of one loop in the middle of the upper face of the fixed ruler. The movable ruler is then said to be hung by a Bifilar Suspension; the effective lengths (l cm.) of the threads supporting it can be taken to be the arithmetic mean of the differences of height above the bench of the upper faces of the fixed and movable rulers at the points where the threads touch them.

It would seem to simplify the adjustment of the lengths of the supporting threads to equality if we made fairly large loops at the ends of two pieces of thread, slipped the movable ruler into the loops and clipped the loose ends of thread into the clips of two retort stands. But if this is done, the effective lengths of the supporting threads for *torsional* oscillations would probably be the lengths of the single threads between the retort-stand clips and the tops of the loops round the lower ruler. For during the whole of a small *torsional* oscillation the face of the lower ruler remains horizontal, as AB in (i), where O is the point of support and P the top of the loop; the friction between thread and ruler prevents P from moving; so the effective length of thread is OP. It is worth while to convince yourself of this by trying it out, and contrasting it with a *pendular* oscillation of the ruler, in which every point moves in a plane parallel to the paper, as in (ii); here AB tilts and the effective length of the supporting thread is from O to AB.

(iii) is an out-of-scale diagram to show what happens when two complete loops are used for a torsional oscillation.

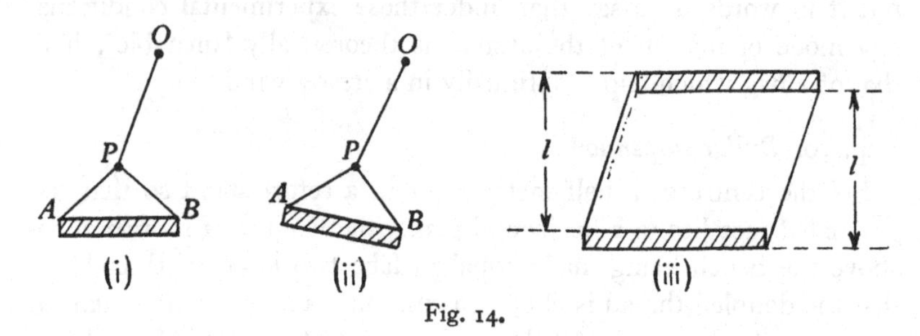

Fig. 14.

Adjust the two loops of thread to pass over or under both rulers at the 15 and 35 cm. graduations, so that $a = 10$; twist the lower ruler through a small angle, keeping its centre as nearly undisturbed as you can; when the ruler is released it will perform torsional oscillations, nearly about a vertical axis through its C.G. The periodic time of these swings will not be affected by any other small oscillations of the ruler, but it is easier to time them if other oscillations are absent. Measure carefully the periodic time (T sec.) of the torsional oscillation.

Repeat with a series of values of $2a$, the centre of the lower ruler being midway between the threads in each case, and plot a graph connecting $1/T$ and a.

The appended note on theory shows that

$$T = \frac{A}{a}\sqrt{l},$$

where A is a constant, so that a graph connecting $1/T$ and a, when l is constant, should be a straight line through the origin.

By measuring T for a series of values of l, with a constant, and plotting the graph connecting T^2 with l, a further check of the theory can be made, for this graph should theoretically be a straight line through the origin.

Note on the Theory

It is shown in books on Mechanics that, if a body is free to rotate about an axis, the moment of the forces acting on it about that axis is equal to the angular acceleration produced by those forces multiplied by a quantity which represents its 'rotational inertia' about that axis; this quantity is usually called its Moment of Inertia about the axis, and is denoted by I. Compare this with Newton's law that if a body, free to move in a straight line, is acted on by a force in that straight line, the force is equal to the linear acceleration multiplied by the mass.

Again, it is shown that if the body moves so that it has an angular acceleration about the axis round which it can rotate, which is proportional to its angular displacement (θ) from a fixed position and of opposite sign to θ, that is, if the angular acceleration is $-A\theta$, where A is a constant, then the body will oscillate with S.H.M. and its periodic time (T) will be $2\pi/\sqrt{A}$. Compare this with the periodic time of a body moving with S.H.M. in a straight line, when the linear acceleration is $-A$ times the displacement.

Suppose now that in a particular case it is proved that the moment of the forces on a body about the axis round which that body can rotate is N times the angular displacement θ (N being a constant) and opposite in sign to θ; that is, suppose that the moment of the forces is $-N\theta$. Then the angular acceleration must be $-N\theta/I$, by the first statement, and N/I takes the place of A in the second statement, so the periodic time of the body will be $\dfrac{2\pi}{\sqrt{(N/I)}}$ or $2\pi\sqrt{(I/N)}$.

The matter can be put briefly in this particular case as follows. By the first statement the moment of the forces is $I\dfrac{d^2\theta}{dt^2}$; we suppose that in this particular case the moment of the forces is $-N\theta$, where N is a constant; hence

$$I\frac{d^2\theta}{dt^2} = -N\theta. \qquad \text{(i)}$$

The relation $$\theta = B \sin\left\{\sqrt{\frac{N}{I}}\,t + \alpha\right\}$$

satisfies the equation (i), where B and α are any constants; this can be shown by differentiating this value of θ twice with respect to t.

Hence θ has the same value at times t, $t+\frac{2\pi}{\sqrt{(N/I)}}$, $t+\frac{4\pi}{\sqrt{(N/I)}}$, etc.; hence the periodic time of the oscillation is $2\pi\sqrt{(I/N)}$.

To apply this general reasoning to this particular experiment we must first calculate the moment of the forces on the movable ruler when displaced through a small angle θ from its equilibrium position.

Suppose that a bar AB is hung horizontally from fixed points P and Q in the same horizontal line by parallel threads of equal lengths l cm., attached to points M and N of the bar at equal distances a cm. from its C.G. C.

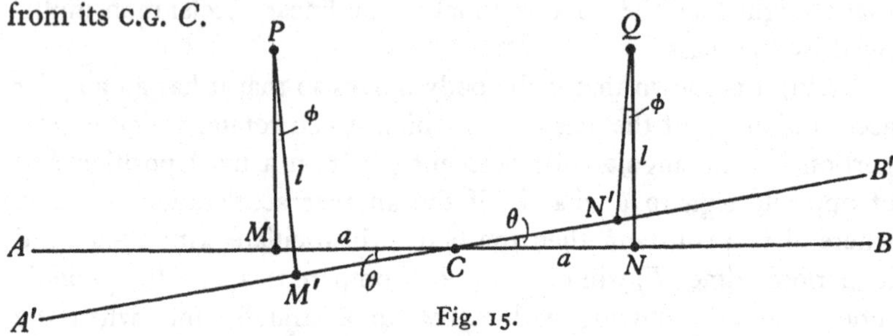

Fig. 15.

Suppose that AB is turned through a small angle θ about a vertical axis through C to a position $A'B'$, so that M and N occupy the positions M' and N'. Suppose that each thread is now inclined at a small angle ϕ to the vertical. θ and ϕ are supposed to be so small that $\sin\theta$ and $\sin\phi$ may be taken as equal to θ and ϕ, measured in radians.

If the mass of AB is m g., the tension of each thread will be $\frac{1}{2}mg$ dynes, and the horizontal component of the tension in PM will be $\frac{1}{2}mg\sin\phi$, or $\frac{1}{2}mg\phi$ dynes, along MM'. So the restoring moment of the forces on the rod about a vertical axis through C will be $2a\frac{mg\phi}{2}$ or $amg\phi$ dynes cm.

Now $MM'=l\phi$, and $MM'=a\theta$, so $\phi=\frac{a}{l}\theta$, and the moment of the forces will be $\frac{a^2mg\theta}{l}$ dynes cm.

Hence, denoting the moment of inertia of the bar about C by I, and applying the results of the foregoing general reasoning, the bar will oscillate with a periodic time

$$T=2\pi\sqrt{\frac{Il}{a^2mg}} \quad \text{or} \quad \frac{2\pi}{a}\sqrt{\frac{Il}{mg}}.$$

Hence I/m for the bar equals $\frac{T^2a^2g}{4\pi^2l}$.

Exp. 11. *Torsional oscillations of a suspended bar*

Exp. 10 dealt with bifilar suspension of the ordinary kind, where the C.G. of the suspended body is midway between the two supporting threads.

If a bar is suspended by a single thread it will not lie horizontally unless the thread passes through the C.G. of the bar, but if it is supported by two threads it will do so for any position of its C.G., provided that that position lies between the threads. The theory of this mode of oscillation is rather complicated, but the experimental verification of the results of that theory is simple, and that is the purpose of this experiment.

Find the scale-reading of the C.G. (C) of a half-metre ruler, by balancing it on a knife-edge. Suspend the ruler, as in Exp. 10, in two vertical loops of thread 20 cm. apart, making $CM=17$ cm. and $CN=3$ cm., C, M and N being as shown in Fig. 16.

Displace the ruler by turning it through a small angle about a vertical axis through C; release it, and measure the periodic time (T sec.) of its torsional oscillations. Do not measure T unless and until you have started the ruler swinging so that C does not move appreciably during the oscillations.

Repeat for a series of positions of M and N, for a range of CM from 17 to 10 cm., MN being 20 cm. in each case; P and Q can remain fixed on the upper ruler.

Calculate corresponding values of $CM \times CN$ and $1/T^2$, and plot a graph connecting them.

If, as is probable, the graph is a straight line through the origin, determine its slope to the axis of $1/T^2$.

The appended note shows that the graph should theoretically be a straight line through the origin, whose slope to the $1/T^2$ axis is $\frac{4\pi^2 Il}{mg}$, where I is the moment of inertia of the ruler about an axis through C perpendicular to its face; I/m may be taken to be one-twelfth of the square of the length of the ruler. Compare the calculated and observed slopes in this case.

Note on the Theory

This is an extension of the note appended to Exp. 10; here C is not midway between M and N, and we suppose that the bar AB is initially rotated through a small angle θ about a vertical axis through O, any point on the bar other than C.

If T_1, T_2 are the tensions in the threads PM and QN, and ϕ_1, ϕ_2 are the small angles between PM and PM' and between QN and QN', and if $CM=x_1$ and $CN=x_2$, since the bar is in equilibrium, m g. being its mass,

$$T_1+T_2=mg,$$

and by taking moments about C

$$T_1x_1=T_2x_2.$$

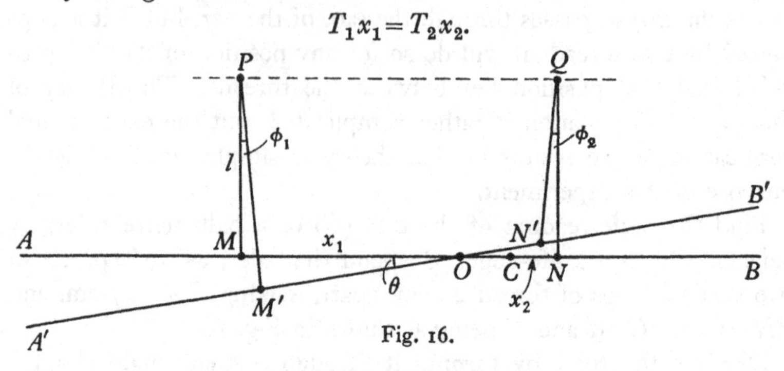

Fig. 16.

Here we have considered only forces in a vertical plane, so that the horizontal forces needed to keep the bar displaced through an angle θ do not come in.

Hence $$T_1=\frac{x_2}{x_1+x_2}mg \quad \text{and} \quad T_2=\frac{x_1}{x_1+x_2}mg.$$

The horizontal components of these forces, very nearly at right angles to AB, along MM' and NN' are $T_1\phi_1$ and $T_2\phi_2$, since ϕ_1 and ϕ_2 are small.

If O is to remain unmoved through the oscillation when the forces at A and B that produced the displacement have been removed, so that O is the fixed centre of oscillation, $T_1\phi_1$ must equal $T_2\phi_2$; if it does not, there will be a broadside resultant on the ruler which will displace O.

But $$MM'=OM\theta \quad \text{and} \quad MM'=l\phi_1$$

and $$NN'=ON\theta \quad \text{and} \quad NN'=l\phi_2.$$

Hence $$\phi_1=\frac{OM}{l}\theta \quad \text{and} \quad \phi_2=\frac{ON}{l}\theta.$$

Substituting for T_1 and ϕ_1 in $T_1\phi_1$, and for T_2 and ϕ_2 in $T_2\phi_2$, since $T_1\phi_1=T_2\phi_2$ we get

$$\frac{x_2 mg}{x_1+x_2}\frac{OM}{l}\theta=\frac{x_1 mg}{x_1+x_2}\frac{ON}{l}\theta$$

or $$x_2\,OM=x_1\,ON \quad \text{or} \quad \frac{x_1}{x_2}=\frac{OM}{ON}.$$

Hence O must be at C, and *the ruler will oscillate steadily only about a vertical axis through its centre of gravity.*

The moment of the horizontal components about C is

$$x_1 T_1 \phi_1 + x_2 T_2 \phi_2.$$

Now we have shown that $OM = x_1$, $ON = x_2$, so $\phi_1 = \frac{x_1}{l}\theta$ and $\phi_2 = \frac{x_2}{l}\theta$. Substituting for T_1, T_2, ϕ_1 and ϕ_2, the moment becomes

$$\frac{x_1 x_2 mg}{x_1 + x_2}\frac{x_1}{l}\theta + \frac{x_1 x_2 mg}{x_1 + x_2}\frac{x_2}{l}\theta \quad \text{or} \quad \frac{x_1 x_2 mg}{l}\theta.$$

Hence, as shown in the note to Exp. 10, the ruler will oscillate torsionally with a periodic time

$$2\pi\sqrt{\frac{I}{(x_1 x_2 mg)/l}} \quad \text{or} \quad 2\pi\sqrt{\frac{Il}{x_1 x_2 mg}},$$

where I is its moment of inertia about a vertical axis through its C.G.

If its periodic time is T sec.

$$x_1 x_2 = \frac{4\pi^2 Il}{mg}\frac{1}{T^2}.$$

Exp. 12. *Moments of inertia*

Use the same set-up as in Exp. 10, but keep the length (l) of the loops of thread, and the distance ($2a$) between them, constant.

Measure the periodic time (T sec.) of torsional oscillation of the bar, and calculate its moment of inertia I from the formula given in the note appended to Exp. 10, $I = \frac{T^2 a^2 mg}{4\pi^2 l}$, using values of a, m and l measured from the apparatus. Compare this with the theoretical value of I for the bar, one-twelfth of its mass multiplied by the sum of the squares of its length and breadth.

Next, investigate the increase in the moment of inertia caused by loading the bar. Put two equal weights, each of mass m_1 g. (say $m_1 = 20$), one on each side of the C.G. of the bar, with their centres equally distant (r_1 cm.) from the C.G.

Measure the periodic times (T sec.) of torsional oscillation of the loaded bar, and deduce from the above formula for I its value in this case; m in the formula is now the sum of the masses of the bar and the two added weights. Subtract from this the moment of inertia of the bar itself, already determined; the result is the rotational inertia (I_1) contributed by the two additional weights r_1 cm. from the C.G.

Repeat for a series of values of r, and plot a graph connecting I_1 and r_1^2.

If, as is probable, this graph is a straight line, find its slope to the axis of r_1^2 and its intercept on the I_1 axis.

The appended note shows that the former should theoretically be the mass of the added weights, $2m_1$, and the latter should be twice the moment of inertia of one of the weights about a vertical axis through its C.G. The latter is so small in comparison with the other moments of inertia involved that no reasonably accurate value can be expected.

The general result of the experiment should show that each component body makes two contributions to the general moment of inertia about an axis, one equal to its moment of inertia about a parallel axis through its C.G., the other equal to its mass multiplied by the square of the distance of its C.G. from the axis.

Note on the Theory

The moment of inertia of a rigid body about an axis is by definition the sum of the mass of every particle of the body multiplied by the square of the distance of that particle from the axis.

It is shown in text-books on Mechanics that if the body is a rectangular bar and the axis is a line through its centre of gravity perpendicular to one of its faces, the moment of inertia is one-twelfth of the mass multiplied by the sum of the squares of the length and breadth of that face; that if it is a circular cylinder and the axis passes through the centre of all the circular cross-sections, the moment of inertia is one-eighth of the mass of the cylinder multiplied by the square of its diameter; and further, that the moment of inertia of a body about an axis which does not pass through its C.G. is equal to the sum of its moment of inertia about a parallel axis through its C.G. and its mass multiplied by the square of the distance of its C.G. from the axis.

Thus the contribution of the two weights in the experiment, if their masses are each m_1 g. and their diameter d_1 cm., should be $2\frac{m_1 d_1^2}{8}+2m_1 r_1^2$. Hence a graph connecting I_1 and r_1^2 should have an intercept on the I_1 axis of $\frac{m_1 d_1^2}{4}$, and a slope to the r_1^2 axis of $2m_1$.

In the weights used in this experiment, d_1 is small compared with r_1, so the intercept will be small.

Exp. 13. *Effects of fluid friction*

The purpose of the following experiment is to check some of the results of a piece of theory which is of considerable importance in predicting the behaviour of a given design of aeroplane, submarine, torpedo or surface vessel, when the friction between its outer surface and the surrounding fluid can be taken to be proportional to the first power of its speed. For very high speeds this does not hold good, but for ordinary speeds it gives a reasonably good approximation to the truth.

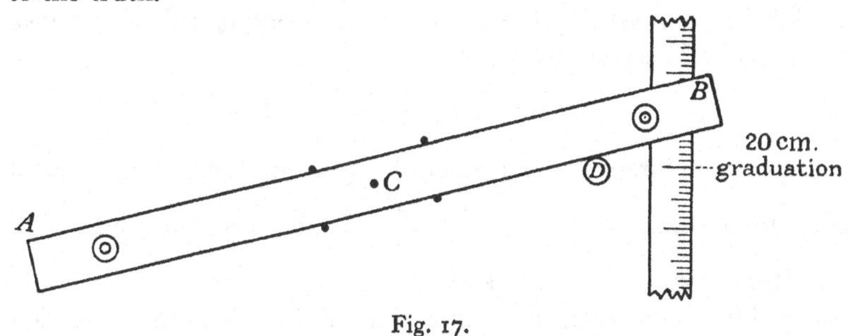

Fig. 17.

The application of this theory enables us to measure the moment of the frictional forces on a body of any shape when oscillating about an axis through its centre of gravity; in this case the body is hung with bifilar suspension, to keep it oscillating.

Use the same set-up as in Exp. 10, making l about 30 cm., but make $2a=4$ cm. about, and put two 20 or 50 g. weights, one near each end of the swinging ruler AB at equal distances from its centre; the lower ruler should be about 1 cm. from the bench.

Find the periodic time, T sec., of torsional oscillations of the ruler AB together with the weights on it, and its moment of inertia I as in Exp. 11 or by calculation of its theoretical value given in the note to Exp. 12.

On the bench, near one end of AB, lay another ruler with its graduated edge outwards, as in Fig. 17; adjust it so that its 20 cm. graduation (say) lies exactly under one edge AB of the ruler when it is at rest. Small angles of swing can then be measured by the displacement (y), along the fixed lowest ruler, of the edge AB from its equilibrium position.

Displace the swinging ruler about a vertical axis through C, so that y is 7 or 8 cm., and put a bottle (D) on the bench to hold it in that position.

Remove the bottle in such a way as not to jerk AB; AB should then take up torsional oscillations about C, disturbed to only a negligible extent by other oscillations.

Observe y when AB is momentarily at rest after making 0, 5, 10, 15, etc., complete swings up to about 60 swings. Denote the value of y after n such swings by y_n.

Plot a graph connecting $\log y_n$ with n. It will probably be a straight line.

The appended note on theory shows that if the moment of the frictional forces resisting the motion of the swinging ruler is μ times its angular velocity at any instant,

$$\log y_n = \log y_0 - 0{\cdot}4343 \frac{\mu T}{2I} n.$$

Hence your graph should theoretically be a straight line, with a slope to the axis of n of $0{\cdot}4343 \frac{\mu T}{2I}$. From this and your graph calculate the value of μ in this case.

It will be seen that the logarithm of the amplitude of the oscillation decreases by a constant amount for each successive swing, since it is proportional to n; this constant is usually called the 'logarithmic decrement'.

Note on the Theory

We have here to extend the theory set out in the note to Exp. 10 to cover cases where the motion is resisted by frictional forces which vary directly with the speed of the body.

Suppose that a body can rotate about a fixed axis, and that when displaced through an angle θ from its equilibrium position it is acted on by external restoring forces whose moment about that axis is $-N\theta$, where N is a constant, and that when its angular velocity is $d\theta/dt$ the moment about the axis of the frictional forces is $-\mu \frac{d\theta}{dt}$.

Then, extending the note to Exp. 9, the equation of motion of the body will be

$$I \frac{d^2\theta}{dt^2} = -\mu \frac{d\theta}{dt} - N\theta. \tag{1}$$

If the damping effect of friction is small, as in this experiment, the relation between θ and t which satisfies (1) is

$$\theta = Be^{-\mu t/2I} \sin (pt + \alpha), \tag{2}$$

where B and α are any constants and $p^2 = \frac{N}{I} - \frac{\mu^2}{4I^2}$.

This can be shown to be the case by differentiating (2) with respect to t and substituting the results in (1).

Suppose we start timing when $\theta=0$, then from (2)

$$0=B\times 1 \sin \alpha,$$

so $\alpha=0$ and the relation becomes

$$\theta=Be^{-\mu t/2I} \sin pt. \qquad (3)$$

The body will therefore be again at the $\theta=0$ position and moving in the same direction when $t=2\pi/p$, $4\pi/p$, $6\pi/p$, etc., so its periodic time $T=2\pi/p$. It will periodically come to rest when $\dfrac{d\theta}{dt}=0$; from (3)

$$\frac{d\theta}{dt}=Bpe^{-\mu t/2I}\cos pt-B\frac{\mu}{2I}e^{-\mu t/2I}\sin pt,$$

so it will momentarily come to rest when $\tan pt=2pI/\mu$. If we assume, as is true in this experiment, that μ is small compared with pI, so that pI/μ is almost infinitely large, then it will come to rest when $pt=\frac{1}{2}\pi$, $\frac{3}{2}\pi$, $\frac{5}{2}\pi$, etc., or when $t=\frac{1}{4}T$, $\frac{3}{4}T$, $\frac{5}{4}T$, etc. If we consider only the maximum extensions on one side of the equilibrium positions, these will occur at times $\frac{1}{4}T$, $\frac{5}{4}T$, $\frac{9}{4}T$, etc.

These maximum extensions will then be $Be^{-\mu T/8I}$, $Be^{-5\mu T/8I}$, $Be^{-9\mu T/8I}$, etc., so they will be in the ratios 1, $e^{-\mu T/2I}$, $e^{-2\mu T/2I}$, $e^{-3\mu T/2I}$, etc. Hence if we measure y_n along a ruler as in the experiment, $y_n=y_0e^{-n\mu T/2I}$.

This is the Exponential Curve, referred to in Exp. 16.

Taking logs of both sides,

$$\log y_n=\log y_0-\frac{n\mu T}{2I}\log e,$$

and since $\log e=0{\cdot}4343$ we get finally

$$\log y_n=\log y_0-\frac{0{\cdot}4343\mu T}{2I}n.$$

Exp. 14. *Measurement of viscosity of water*

The following experiment enables us to measure the viscosity of cold water, using only very simple apparatus.

Suitable apparatus for this experiment consists of a 1 gal. paint or distemper can, about 17 cm. diameter, or any vessel of comparable diameter, with a brass tube about 1 cm. external diameter and 3 cm. long soldered into the side near the bottom of the tin, forming a horizontal nozzle.

This tin is to be supported about 25 cm. above the bench; a length of about 25 cm. of thick-walled capillary glass tubing, with a bore of about 1 mm. diameter, is to be connected to the nozzle by rubber tubing 15–20 cm. long; a spring or screw clip on this tubing is convenient but not necessary. The tin is to be filled nearly to the brim with cold water, of measured temperature (t° C.).

All air-bubbles should be got out of the rubber tubing by pinching it with the fingers while water is flowing through; the capillary tube is then fixed horizontally by a retort stand, first at about the level of the nozzle.

The height of the free surface of water in the tin above the bench is to be measured, and also the height above the bench of the bore of the capillary; the difference is the hydrostatic head of water above the inlet, or the pressure difference between the ends of the tube when the outlet is open to the air.

A measuring cylinder is then put under the outlet, and the number (T) of seconds required for the discharge to amount to a fixed volume, say 25 c.c., is measured. To facilitate getting an accurate result a piece of wetted string may be tied round the capillary tube close to its outlet and allowed to hang down into the measuring cylinder, touching its side but not reaching below the 25 c.c. graduation; the flow into the cylinder is then steady and the meniscus is not disturbed by periodic drops.

This is to be repeated for four or five different heights of the tube above the bench, the capillary tube being horizontal in each case; the range of heights should be the full amount allowed by the rubber tube without much kinking.

A graph connecting $1/T$ (which is proportional to the rate of flow through the tube) and the difference of pressure h between the ends of the tube, should be plotted.

If, as is probable, this is a straight line through the origin you will have shown experimentally that *the rate of discharge from a horizontal capillary is directly proportional to the difference of pressure between its ends.*

Viscosity. The appended note on theory shows that in this case the viscosity of water at this temperature is $\frac{\pi p r^4 T}{8lV}$, where $p=hg$ if the density is 1 g. per c.c., l cm. is the length and r cm. the radius of bore of the capillary, and V c.c. of water are discharged in T sec.

r should be determined by weighing a thread of mercury of

measured length in the capillary, and l by measuring the length of the capillary tube. The slope of the graph gives the ratio of h to $1/T$, and V is known. Hence we can calculate the value of the viscosity of water at $t°$ C. The result should be compared with the recognised value by plotting a graph of the values given in tables of physical constants for various temperatures. Kaye and Laby quote them as 0·01522, 0·01311, 0·01142, 1·01006 and 0·00893 at 5, 10, 15, 20 and 25° C. respectively. From this graph you can read off the recognised value at $t°$ C.

Note on the Theory

The viscosity of a fluid can be defined as follows: 'If two parallel planes are at unit distance apart in a fluid, and one of them is moving in its own plane with unit velocity relatively to the other plane, then the tangential force exerted per unit area on each of the planes is equal to the viscosity.'

In this definition there are many unexpressed conditions; for example, that the fluid 'wets' the planes; to deal with them a chapter in a text-book of Hydrodynamics is required. But there is, in addition, a limitation which this definition shares with most other definitions; it does not pretend to state physical laws. For instance, the definition of unit quantity of electricity in electrostatics is 'that quantity which placed 1 cm. distance from an equal like quantity repels it with a force of 1 dyne', and it gives no clue to the force of repulsion between such quantities at a distance n cm. apart.

So, in the definition of viscosity, we have no clue to the tangential forces when the planes are l cm. apart, or when the relative velocities are v units.

But if we make certain hypotheses about the effect of such changes in l, v, etc., we can work out theoretically the effects of viscosity in restricting the rate of flow through a capillary tube; we can then, by an experiment such as this, test the truth of this result in a variety of cases, and thus confirm or disprove the truth of the hypotheses on which the result was founded.

In this case the 'result' is that the viscosity of water flowing steadily through a horizontal capillary tube of length l cm. and radius r cm., when a pressure difference of p dynes per sq.cm. is maintained between its ends, and V c.c. is delivered in t sec., is $\frac{\pi p r^4 t}{8lV}$. We can check this experimentally by measuring V, as in this experiment, when p, r, t and l have a series of values; here p alone has been varied. If we get a constant value for the viscosity in all these cases, we confirm the hypotheses. The two numerical factors π and 8 come direct from the definition, and we cannot confirm these except by checking the mathematical working; but it is necessary to take these numerical factors into account if we want to compare the value we find for the viscosity of water at a certain temperature with the value given in a book of Physical Constants.

It is of some importance to realise that we can experimentally confirm the physical hypotheses by which the expression $\frac{\pi p r^4 t}{8lV}$ was obtained for the viscosity, without referring to any book of Physical Constants, merely by finding whether $\frac{pr^4 t}{lV}$ remains constant when a series of experimental values of p, etc., is substituted in it. Agreement between our value of the viscosity, when account is taken of the numerical factors π and 8, and the recognised value furnishes only an overall check of our experimental methods, accuracy of observation and numerical working.

Exp. 15. *Steady flow through a siphon*

If a siphon is used to draw liquid from a large vessel, the rate of flow soon becomes steady; this experiment is designed to indicate how you can discover the way in which this rate of flow depends on one or two of the factors in the set-up.

Use the same set-up as in Exp. 14, except that it is convenient to have a longer rubber tube, say 40 cm.

Fix the capillary tube at various inclinations, including vertical; the rubber tube may be disposed in any way, e.g. parts of it may be higher than the free surface of the water in the tin so that it and the capillary tube form a siphon.

In each case measure h cm., the difference in heights, above the bench, of the free surface and of the outlet of the tube. Measure also the number (T) of seconds required for the discharge from the capillary to amount to a fixed volume, say 25 c.c. Plot a graph connecting $1/T$ with h.

If, as is probable, your graph is a straight line through the origin you will have shown experimentally that *the rate of steady flow through a capillary is directly proportional to the head of liquid above its outlet.*

Exp. 14 was a particular case of this, designed to measure the viscosity of water as a physical constant. In this more general case we have to consider separately what we may call the total effective driving force, and the viscous resistance to flow.

Each part of the discharge tube, whether rubber or glass capillary, makes its contribution to the viscous resistance, unaffected by its inclination to the vertical; the discharge tube as a whole offers a constant fluid resistance, however it is disposed. On the other hand, the water in each part of the tubing, as well as in the tank, contributes to the effective driving force by adding (algebraically) to the head; if the flow in any piece of the discharge tube is upwards the contribution of that piece to the total head is negative; in fact, the contribution of any piece is its resolved component length in a downward direction, so that the total driving force due to the water in tank and tube is the depth of the outlet below the free surface in the tank.

For example, if we take a siphon consisting of an inverted U of capillary glass tubing, and put one leg in the tank, and move the siphon up and down, the rate of flow will vary in proportion to the depth of the outlet below the free surface, since that is the way in which the total driving force changes, while the viscous resistance does not change.

But if we keep the siphon fixed, and break off part of the leg below the surface in the tank, the rate of flow will increase because, though the driving force is unchanged, the viscous resistance has been decreased.

Exp. 16. *Emptying a vessel of liquid*

It is obviously difficult to measure directly the speed of a body at some assigned instant when that speed is changing from instant to instant. It can be done indirectly by means of tangents to a graph connecting displacements and time, as in the usual elementary treatment of uniformly accelerated motion, but if an algebraical relation can be discovered between them it is possible that a much simpler and more accurate method can be designed. This experiment illustrates this possibility, in a case where the acceleration is not uniform.

Fix a 25 c.c. burette vertically in a stand. By means of a rubber tube about 10 cm. long, fitted with a spring or screw clip, connect the nozzle of the burette to a piece of thick-walled capillary glass tubing about 25 cm. long with a bore of diameter about 1 mm., fixed roughly horizontally in a stand. As explained in Exp. 14 it is advisable to tie a wetted piece of string round this tube near its outlet, to steady the discharge.

Fill the burette with cold water above the 0 c.c. graduation. Open the clip and by means of a time-piece with a seconds hand note the times at which the meniscus in the burette passes the 0, 5, 7, 10, 12, etc., ..., 25 c.c. graduations; deduce the number (t) of seconds which have elapsed in each case since passing the 0 c.c. graduation.

Measure the head (h cm.) of each of these graduations above the outlet of the capillary tube.

Plot a smooth graph connecting h as ordinate and t as abscissae.

The next step can be taken by either of the two following methods; the first (A) requires little or no knowledge of mathematics, but is more laborious and less accurate than the second (B) which involves some knowledge of the differential calculus.

(A) Draw tangents to the curve at a series of points and calculate the slopes of these tangents to the axis of t. The slope of the tangent at any point is the time-rate of decrease of h, when h has the value corresponding to that point. Plot a graph connecting these slopes with h; if, as is probable, this is a straight line through the origin, you will have shown experimentally that the rate of decrease of h is directly proportional to h.

(B) Complete your table of corresponding values of h and t by a column showing $\log h$, and plot a graph connecting $\log h$ and t. If, as is probable, this is a straight line, which does not pass through the origin, the appended note on theory shows that the time-rate of decrease of h must be directly proportional to h.

(A) and (B) are alternative methods of arriving at the characteristic property of your first graph (which is called an Exponential Curve), that the rate of decrease of the ordinate with respect to the abscissa is directly proportional to the ordinate. This curve often occurs in physics; for instance, in the rate at which a charged condenser discharges through a high resistance, or in Newton's law of cooling, or in the rate at which the needle of a dead-beat instrument approaches its final position, or the rate at which the amplitude of a simple pendulum dies away (see Exp. 13). This particular experiment shows that the same law governs the rate at which a bath empties itself through the waste.

If r cm. is the radius of the burette, the rate of decrease in c.c. per sec. of the volume of water in the burette (i.e. the rate of discharge through the capillary tube) is $\pi r^2 \times$ rate of decrease of h; so the experiment shows that *the rate of discharge through the capillary at any instant is directly proportional to the head of water above it at that instant.*

Exp. 15 showed that the same thing was true when the flow was almost exactly steady, because the area of the surface of water in the tank was so large that its drop during the experiment was negligibly small; we have now shown that it is also true at all instants when the rate of flow is not steady but decreasing. In this latter case we have to determine the rate of discharge from observations of the movement of the surface of water in the burette; we cannot do so by collecting the water discharged in a finite time, as we did in Exp. 15, since the rate of discharge now varies from moment to moment, instead of being steady as in that experiment.

Note on the Theory

Suppose that observations show that two variables h and t are connected, over a certain range of values of h, by the relation

$$\log_{10} h = at + b, \tag{1}$$

where a and b are constants.

Putting $t=0$, $b=\log_{10} h_0$, where h_0 is the value which h would have when $t=0$ if the relation (1) held good at that time; $\log_{10} h_0$ is in fact the intercept on the axis of $\log_{10} h$, whether or no (1) holds good when $t=0$. Hence

$$\log_{10}\left(\frac{h}{h_0}\right) = at. \tag{2}$$

Now, if n is any number,

$$\log_{10} n = \log_{10} e \times \log_e n$$
$$= 0{\cdot}43429 \log_e n,$$

where $e = 2{\cdot}7183\ldots$. So (2) becomes

$$0{\cdot}43429 \log_e\left(\frac{h}{h_0}\right) = at \quad \text{or} \quad \log_e\left(\frac{h}{h_0}\right) = 2{\cdot}3026at.$$

Hence, from the definition of a logarithm,

$$\frac{h}{h_0} = e^{2\cdot3026at} \quad \text{or} \quad h = h_0 e^{2\cdot3026at}, \tag{3}$$

and the graph connecting h and t is called an Exponential Curve.

It is shown in books on the differential calculus that $de^{At}/dt = Ae^{At}$, where A is any constant. Hence, differentiating both sides of (3) with respect to t, we get

$$\frac{dh}{dt} = h_0 2{\cdot}3026ae^{2{\cdot}3026at}$$

or, by (3), $$= 2{\cdot}3026ah,$$

expressed in words, the time-rate of change of h is directly proportional to h.

If the graph connecting $\log_{10}h$ and t is a straight line inclined to the axis of t in such a way that a is negative, then the time-rate of *decrease* of h (or $-dh/dt$) is directly proportional to h.

It will be seen that h_0 has disappeared before reaching the final result.

Exp. 17. *Measurement of surface tension*

Take a thermometer tube, a thick-walled glass capillary tube with a uniform bore not more than 1 mm. diameter, and about 14 cm. long. Tie tightly round it a single turn of cotton, about 1 cm. from one end.

Measure the radius, r cm., of the bore by sucking mercury into it, so that the mercury occupies a measured length l cm. of the tube; find the mass m g. of this mercury and deduce r from the relation $\pi r^2 l\ 13{\cdot}6 = m$.

Clean the bore of the capillary tube, so that the liquid whose surface tension has to be measured 'wets' it and makes a zero angle with the glass where the meniscus meets it, which in the experiment will be at the level of the cotton. If the liquid is methylated spirits, which will be used in this experiment, this can generally be done well enough by dipping the end with the cotton on it into water and then into the methylated spirits, and in each case sucking the liquid up several times. If the liquid is benzene, turpentine, paraffin or petrol, these must clearly be the last liquids with which the tube is flushed. If the liquid is water, more elaborate methods of cleaning are necessary, since the surface of water is very easily contaminated; it is the surface of the liquid rather than that of the glass which is important.

Set up apparatus as in Fig. 18, where A is a 50 c.c. measuring cylinder or a beaker containing the methylated spirits which can be brought up round the capillary tube and supported there by blocks.

The tube is to be fixed vertically in a retort stand, cotton end downwards, at such a height above the bench that the measuring cylinder will slide under it.

B is a simple form of cathetometer, most easily made by taking a board, say, 10 × 30 cm., and fixing a half-metre ruler to it, by means of bits of wood fixed at top and bottom of the board by single screws, the other ends of the bits of wood being packed out from the board, possibly by another ruler. The former ruler is to be set back from the edge of the board so that a set square can slide up and down the ruler; the ruler should be adjusted before tightening the

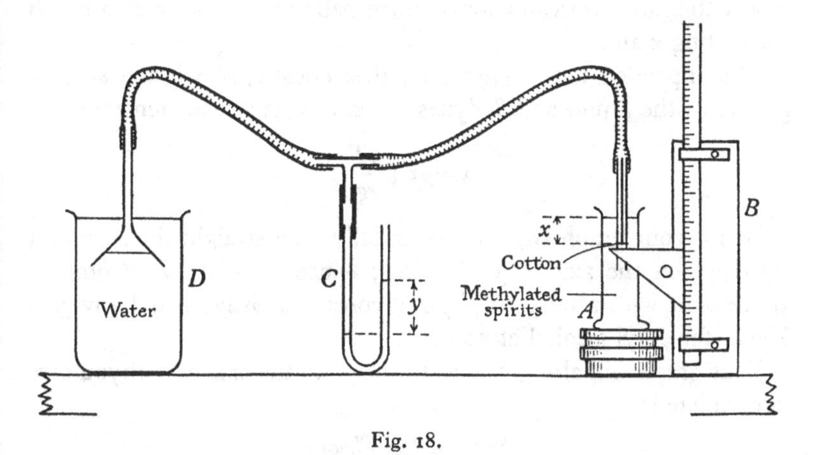

Fig. 18.

screws so that when the bottom of the board sits squarely on the bench the top edge of the set square is shown by a spirit level to be horizontal. These precautions are necessary because the accuracy of the result depends mainly on measuring heights of liquids above the bench and an inaccuracy of 1 mm. in them may make a 10% error in the result.

By means of rubber tubes and a glass T tube connect the top of the capillary tube to a water manometer *C* and to an inverted funnel, held by a retort stand in a vessel of water *D*. It is advisable to lash the rubber tubes to the glass tubes, to make them pressure tight; readings can then be taken at leisure.

Before surrounding the capillary tube with the liquid, take the cathetometer scale-reading of the cotton. Bring up the cylinder and immerse the cotton, roughly 1 cm. below the free surface of the liquid, and support the cylinder by suitable blocks in this position.

Take the cathetometer scale-reading of the free surface in A, and deduce, x cm., the head of methylated spirits above the cotton.

Lower the funnel in D until the increased air pressure in the apparatus brings the meniscus in the capillary nearly down to the cotton; fix the funnel in a retort stand, and add enough water to D to bring the meniscus exactly to the cotton.

Take the cathetometer scale-readings of the water in the manometer and deduce the head, y cm., of water which corresponds to the excess pressure, above atmospheric, of the air in the apparatus.

Next, raise the cylinder A still further and determine x and y; repeat this for a series of four or five pairs of values. Plot a graph connecting x and y.

The appended note shows that, theoretically, if ρ is the specific gravity of the liquid and T dynes per cm. is its surface tension

$$y = \rho x + \frac{2T}{rg}.$$

Hence your graph should theoretically be a straight line, with an intercept on the axis of y of $2T/rg$; hence calculate T. Compare your result with a table of physical constants; Kaye and Laby give $T=22$ for ethyl alcohol at 20° C.

Your graph will also give ρ as the slope to the axis of x, if you care to calculate it.

Note on the Theory

Text-books on Hydrostatics show that in a capillary tube the difference of pressure between the gas on the concave side of the meniscus of a liquid which wets the tube so that the angle of contact with the glass is zero, and the liquid at the lowest point of the meniscus, is $2T/r$ dynes per sq.cm.

In the experiment, the excess of air pressure inside the capillary over atmospheric pressure is yg dynes per sq.cm., and the excess of the pressure in the methylated spirits at the level of the cotton over atmospheric pressure is $x\rho g$ dynes per sq.cm.; hence

$$yg = \frac{2T}{r} + x\rho g \quad \text{or} \quad y = \rho x + \frac{2T}{rg}.$$

PART II. LIGHT

Exp. 18. *Critical angle*

Elementary text-books on Light describe a very elegant experiment, in which a parallel-sided 'plate' of air is immersed vertically in a vessel of liquid, and rotated about a vertical axis, whereby the critical angle from the liquid to air can be measured. This set-up is not found suitable for the very precise measurements of refractive indices needed in industry, and the principles on which these measurements are usually made are illustrated in this experiment, which is rough because only the simplest possible apparatus is used.

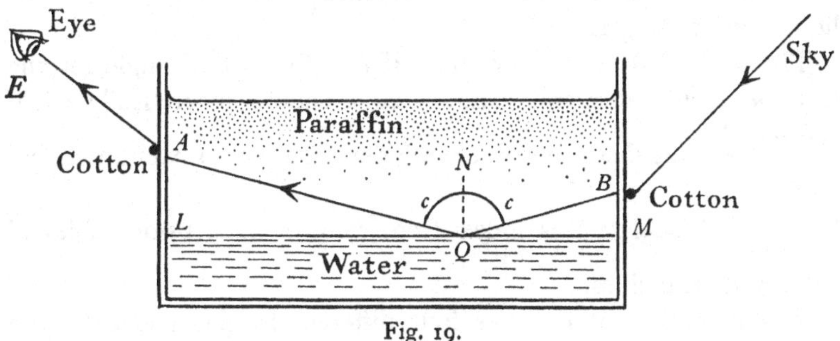

Fig. 19.

Take a wide beaker, or (preferably) a bottle, such as a pickle bottle, with two vertical parallel faces (*AL* and *BM* in Fig. 19) 7 or 8 cm. apart. Pour water into it to a depth of about 5 cm.; add clean paraffin oil to an additional depth of about half *LM*. Tie a piece of black cotton tightly round it, about *LM*/4 cm. above the surface of the water; *A* and *B* are the points where this cotton cuts the plane of the paper.

Stand the bottle near a window, in such a position that, to an eye looking at it as in the figure, but with *E* some 4 ft. from the bottle and not much higher than the surfaces of the liquids, the interface *LQM* appears to be brightly and uniformly lit up; avoid any horizontal window bar which might interfere with this result. The interface will appear to be much brighter than the upper surface of the paraffin, because the light that falls on the former is totally internally reflected towards the eye, while a good deal of the light falling on the latter passes down into the paraffin and does not reach the eye.

Gradually raise the eye, keeping it 3 or 4 ft. from the bottle; at some height it will see a dark shadow, with a diffusely tinted edge, spreading across the interface from the near side. From higher positions of the eye, the interface will appear no brighter than the upper surface of the paraffin.

Repeat this procedure with the eye a few inches from the bottle, as in the figure. Denoting by *Q* the edge of the shadow, adjust the cotton by sliding *A* or *B*, or both, up or down so that when the edge of the shadow appears in line with *A* the reflection of *B* in the interface is brought into the same line; then *A*, *Q* and *B* will appear to the eye at *E* to be in one line, as in the figure.

With a pair of dividers measure *AL* and *BM*, locating *L* and *M* by looking along the interface underneath the curves where the liquids meet the glass.

The appended note shows that, if c is the critical angle for the face between water and paraffin, $\cot c$ should theoretically equal $\frac{AL+BM}{LM}$, and that the refractive index from air to paraffin should be $\frac{1{\cdot}33}{\sin c}$. Calculate this index from your results, using tables of cotangents and sines.

Repeat with *A* at three or four different heights above *L*; you should theoretically get the same value for the refractive index in each case. Calculate the arithmetic mean of your results and compare it with the value 1·44 given in books of Physical Constants.

This method provides very cheaply an optically plane surface of large area, and with such a surface the phenomena can easily be studied. But it works only when the liquid whose refractive index is required is less mechanically dense, but optically denser, than the liquid of known refractive index, and where these liquids do not mix. Liquids, other than paraffin, that fulfil these conditions when used with water, are turpentine and various oils, such as olive oil.

Standard Refractometers, with which the refractive index of any liquid can be measured commercially to a high degree of accuracy, are based on the same optical principles; they use an optically plane surface of a piece of glass whose refractive index is accurately known, as well as telescopes, vernier scales of angles, and so on. But the basic idea can be illustrated with a rectangular block of glass, as ordinarily found in school laboratories, as follows.

Spill some of the liquid (methylated spirits, for example) on a board, and put on it the narrowest, shortest face of a rectangular glass block (whose refractive index ${}_a\mu_g$ has been measured by one of the ordinary methods, such as the parallax method) round which a piece of cotton has been tied, as above. The glass then takes the place of the paraffin in the figure, with its largest face parallel to the plane of the paper; the methylated spirits takes the place of the water. Proceed as before.

Then the required μ is calculated from $\mu = {}_a\mu_g \sin c$, where $\cot c = \frac{AL+BM}{LM}$.

Finally, it is worth while to substitute paraffin oil for the methylated spirits, and try whether the experiment is now workable.

Note on the Theory

Referring to Fig. 19, if NQA is the critical angle (c) for, say, violet light, for the given pair of media, a ray of violet light which enters the eye at E from a point on QM to the right of Q, will have struck QM at an angle greater than the critical angle, and will therefore have been totally reflected; a violet ray which enters the eye from a point on LQ will have struck LM at an angle less than the critical angle, and so will be only partly reflected; the part LQ will therefore appear darker than the part QM.

Since the critical angle for red light is greater than for violet light, the point Q will be shifted to the right, to Q' say, if red light is used instead of violet and if E is in the same place. Hence if white light is used, there will be total reflection of the violet from the band QQ', but the red light reflected from QQ' will be weakened; the band QQ' will therefore have a violet tinge compared with the white appearance of the $Q'M$ part. So the edge of the shadow over LQ will be tinted.

We see from the figure that $\cot c = \frac{AL}{LQ} = \frac{BM}{QM}$, so that both equal $\frac{AL+BM}{LQ+QM}$, or $\cot c = \frac{AL+BM}{LM}$.

If the angle of incidence in the first medium, whose refractive index is μ_1, is ϕ_1, and the angle of refraction in the second medium, of refractive index μ_2, is ϕ_2, then $\mu_1 \sin \phi_1 = \mu_2 \sin \phi_2$; hence if $\phi_1 = c$ and $\phi_2 = 90^\circ$, $\mu_1 \sin c = \mu_2$. In the first part of the experiment this gives for the refractive index of paraffin $\frac{{}_a\mu_g}{\sin c}$ or $\frac{1 \cdot 33}{\sin c}$.

Exp. 19. *Wave fronts*

Take a rectangular glass block, about twice as long as it is broad ($4\frac{1}{2} \times 2\frac{1}{2}$ in. would be satisfactory), put it on a sheet of paper and by means of pins trace the paths of rays proceeding from a pin at a corner A of the block and emerging from the glass. Disregard rays which are internally reflected. About six rays through the long face and three through the short face will suffice if they are well spread.

Deduce μ for the glass from one of these broken rays, hence calculate where the ray from A, which will graze the face BC on emergence, emerges from BC, and draw this broken ray.

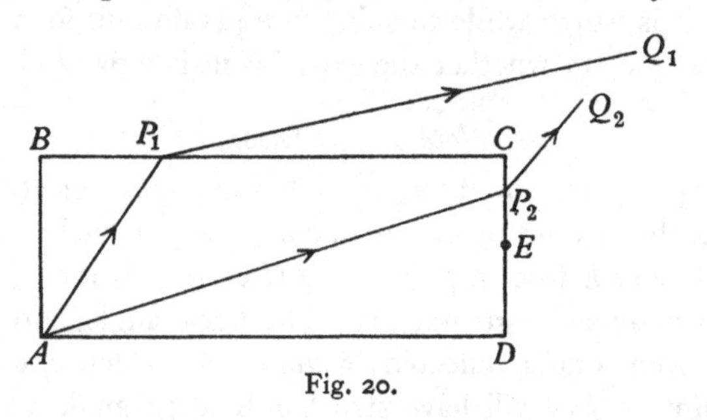

Fig. 20.

Assuming that light travels μ times as fast in air as in glass, if v is the velocity of light in air, and therefore v/μ in glass, the time occupied in travelling along a ray APQ is $\frac{AP}{v/\mu} + \frac{PQ}{v}$ or $\frac{1}{v}(\mu AP + PQ)$. Hence if light starts from A at a given instant and travels along two rays AP_1Q_1 and AP_2Q_2, it will reach Q_1 and Q_2 simultaneously if $\mu AP_1 + P_1Q_1 = \mu AP_2 + P_2Q_2$; Q_1 and Q_2 are then points on the same wave front.

We can therefore mark, on all the rays which have been traced, the points at which they are cut by the wave front which passes through any given point. Take, for example, the wave which has passed along AC in glass and then moved 3 cm. farther in air; we have to find the point on each ray APQ for which

$$\mu AP + PQ = \mu AC + 3, \quad \text{or} \quad PQ = \mu(AC - AP) + 3.$$

Taking μ as 1·5, find Q in this way on each of the traced rays, and sketch in the two smooth curves through these points. Theory shows that all the rays should cut the wave front at right angles, and this can be checked from the diagram.

Similarly, find the wave front which passes through E, the mid-point of CD; this will consist of three curves; the part inside the glass must clearly be a circle with centre A and radius AE.

Exp. 20. *Planning for accuracy*

This experiment illustrates the ways in which theory can help in designing a method of measuring a physical constant to the highest degree of accuracy that simple equipment allows; the constant in this case is the focal length of a convergent lens.

Outline of method. In determining f by means of the theoretical relation $\frac{1}{f} = \frac{1}{u} + \frac{1}{v}$ we have first to locate the image Q_0 of some point P_0 produced by a thin co-axial pencil of rays, and then to measure the distances u and v of P_0 and Q_0 from certain fixed points H_1 and H_2 whose positions relative to the faces of the lens are given by theory.

The most serious errors in determining f arise in the location of Q_0. Theoretically, Q_0 is defined as the intersection of two straight lines which are inclined to one another at an indefinitely small angle, and it is hopeless to base any practical method directly on that definition. Indirectly, theory shows that the definition leads to the production of a moderately sharp image of a bright object, and a commonly used practical method depends on this; but the results are comparatively inaccurate. The ordinary parallax method (see Appendix A) gives tolerable accuracy for a long focus lens, but it is liable to large percentage errors if f does not exceed 10 or 15 cm. In such cases another method is available; it can best be understood by carrying out the following procedure, which is set out in elaborate detail in order to show how minor experimental errors can, and should, be avoided by experimental precautions, based in some cases on theoretical reasoning. The main feature of this method is the use of a straight line which can be shown theoretically to cross the optic axis at Q_0 at a large instead of an indefinitely small angle. It is obvious enough that this is desirable, if theory will provide us with the straight line, and with a practical method of finding points on it; the appended note on theory deals with this side of the matter.

Apparatus required. Assume that we have a lens of focal length about 10 cm. and diameter about 5 cm. and that its faces are parts of spheres of roughly equal radius. Assume that the lens can be mounted in a lens holder which can slide along a base-board or rod, furnished with a metre scale, and that we have an object pin and a search pin (it is better to use needles) mounted in separate blocks which can stand anywhere on the base-board or rod. We shall need also a set square, and a rod with pointed ends such as a knitting-needle, which can be supported so as to lie along the optic axis of the lens (*PCX* in Fig. 21*a*, which is a *horizontal* cross-section of the apparatus, *P* and *X* being the points of the object and search pins).

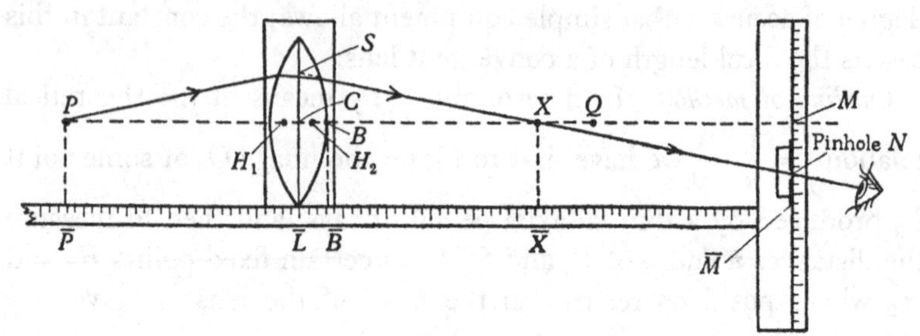

Fig. 21*a*.

We shall also need a cross board fixed to the base-board, carrying a millimetre scale perpendicular to the main scale, along which can slide a block carrying a vertical piece of cardboard with a pin-hole *N* in it; a thin 'fiducial mark' is made on the cardboard, by which changes in the position of *N* can be read on the cross scale.

Measure roughly the focal length (*f* cm.) of the lens by focusing a distant bright object on a screen. Put a spot of ink at the centre, *B*, of one face; this can be done most conveniently by laying the lens on a piece of paper, drawing a line round it, finding and marking the centre of this circle, putting the lens back on it and marking the face.

Set the lens in its holder and make a thin fiducial mark (*L*) on the holder, roughly opposite the mid-point (*C*) of the lens. Arrange a brightly illuminated white card at the far end of the bench, to serve as a background to pins, etc.

Adjustment of apparatus. If the lens is accurately set in its holder, the point *B* will move along the optic axis when the holder slides along the bench; we have now to get the points of the object and search pins, and the pin-hole, at the same level as the ink mark *B* so that all of them can be got into the optic axis.

Put the lens near the far end of the bench and the object pin very close to it, between it and the eye; adjust the height of the pin so that its point seems to coincide with *B* when viewed through the pin-hole. Slide the object pin nearly to the pin-hole, and move the cardboard vertically on its block until the point of the pin again seems to coincide with B when viewed through the pin-hole, moved broadside as necessary. Substitute the search pin for the object pin and adjust the height of its point to coincide with B when viewed through the pin-hole in its new position on the block.

These adjustments can be checked by sliding the lens towards the pin-hole; there should be no relative motion of B and the point of the pin during this motion. Note the cross-scale reading $\overline{N}_0$ which will show when the pin-hole is in the optic axis.

Remove X and move the lens until L is exactly at a graduation about 55 cm. from $\overline{M}$ in Fig. 21 *a*; the lens can then be replaced if accidentally disturbed. Set up the object pin about $2f$ cm. behind the lens and adjust it sideways until the point P of the pin seems to coincide with B when viewed through the pin-hole set at $\overline{N}_0$; P is then in the optic axis.

Observations to be made. The ordinary lens-formula $\frac{1}{u}+\frac{1}{v}=\frac{1}{f}$ is exactly true only when the lens is infinitely thin, if u and v and f are measured as usual from the faces of the lens; it is also true for a lens of finite thickness t cm. if u and v and f are measured from certain Principal Points (H_1 and H_2). Theory shows (see Appendix B) that while the distance of these principal points from the faces A and B of the lens depend in the general case on t, f, μ and the radii of curvature of the faces, the distance H_1H_2 between them is simply $\frac{\mu-1}{\mu}t$, which becomes $\frac{1}{3}t$ if $\mu=\frac{3}{2}$, and only an inappreciable error is caused here by using that value for μ.

Again, theory shows (see appended note) that if u exactly or nearly equals v, any small error in the location of H_1 will have no appreciable effect on the calculated value of f provided that the length H_1H_2 is correct. If the curvatures of the two faces of the lens are exactly the same, the distances of H_1 and H_2 from the two faces will by symmetry be exactly equal to $t/2\mu$, so that even if the curvatures are not equal, only a small error will result from treating these distances as $\frac{1}{3}t$, and this error will not affect the value of f. Hence the following method of locating H_1 and H_2 will be satisfactory; measure the length of the rod, set it approximately in the optic axis with one end touching the face A, find with the help of the set square the scale-reading of the other end and deduce the scale-reading $\overline{A}$. Similarly find $\bar{B}$, and deduce t. Then $\bar{C}$, the scale-reading of the mid-point C of AB, is $\frac{1}{2}t$ from $\bar{A}$, $\bar{H}_1$ is $\frac{1}{3}t$ from $\bar{A}$, and $\bar{H}_2$ is $\frac{1}{3}t$ from $\bar{B}$, as in Fig. 21*a*. Record these values of $\bar{C}$, $\bar{H}_1$ and $\bar{H}_2$ for use later.

Put the object pin in such a position P that PH_1 is nearly equal to $2f$; adjust the object pin sideways until it coincides with the ink-mark when viewed through the pin-hole at $\bar{N}_0$; determine carefully the value of $\bar{P}$, deduce $\bar{P}\bar{H}_1$, or u, and record it for use later.

Set up the search pin X in the optic axis, at about $(2f-1{\cdot}5)$ cm. from H_2. Adjust X broadside with great care, until the point of the image of P seems exactly to coincide with the point of X when viewed through the pin-hole at $\bar{N}_0$; X and P and the ink-mark B will then appear to coincide. Determine carefully the scale-reading $\bar{X}$.

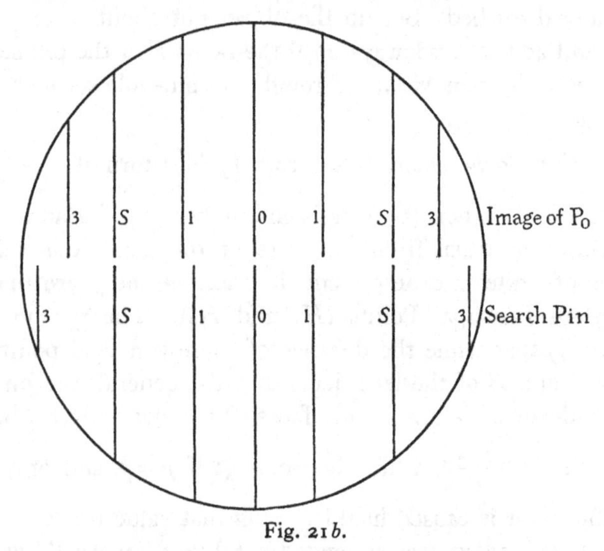

Fig. 21*b*.

On sliding the pin-hole either to left or right of its central position $\bar{N}_0$, P and X should appear to separate slightly, with P leading, then to coincide again, then to separate with X leading towards the edge of the lens. Our object is to get the exact position $\bar{N}$ of the pin-hole when this second coincidence takes place, as shown at S in Fig. 21*b*; the figure also shows successive appearances of X and P as the pin-hole moves broadside.

Take the mean of four or five settings of $\bar{N}_1$ on one side (estimating tenths of a millimetre), and of four or five settings of $\bar{N}_2$ on the other side, of the centre of the cross-slide, and calculate the mean value of $\bar{N}_1\bar{N}_2$. From Fig. 21*a* it will be seen that if y denotes the distance from the optic axis of S, the point at which the central plane of the lens is cut by the emergent ray SX produced backwards, then

$$y=\frac{\overline{N_1N_2}}{2}\,\frac{\overline{CX}}{\overline{XM}}. \tag{1}$$

Hence calculate y. These values of y and $\bar{X}$ serve to fix one of the oblique rays which started from P and passed through X; the pin-hole and the pin X have been used like the back and foresights of a gun for this purpose.

Now theory (see appended note) shows that y and $\bar{X}$ are connected by a simple relation. The distance of X from Q in Fig. 21 *a*, the focus conjugate to P for a thin co-axial pencil, is theoretically proportional to the square of y, or $XQ=k\,y^2$, where k is a constant and Q is some fixed point.

Hence, a graph connecting y^2 and $\bar{X}$ should be a straight line, and when $y^2=0$ in this graph, X should be at Q since the ray SX is then co-axial; hence the intersection of the graph with the axis of $\bar{X}$ should give the scale-reading $\bar{Q}$ of Q.

If then we move X through a series of about five or six steps of 2 or 3 mm. each, and determine as above the corresponding values of y^2, we can plot this graph. The positions of X should extend over as large a range as is practicable, but values of y less than about 1·5 cm. will not give reliable results.

This method of locating a focus may be called the 'Indirect Parallax Method'.

Reduction of observations. Theoretically, this graph should be a straight line; if in practice it is curved, either the lens may be (and sometimes is) slightly faulty in having different focal lengths at different distances from its centre, or else the observations or calculations have been faulty.

In the former case, if a tangent is drawn to the graph at any point y_1^2 the corresponding value of Q will give the focal length of a lens, the whole of whose faces have the same curvature as the zone of the actual lens which is y_1 from the centre.

In the latter case it is usual to get a graph in which one part at least is very nearly straight, though all the points may not lie closely on a straight line; in such an event a really accurate result may be attained by calculating the position of Q by the method of Introduction, § 5, using the graph only to indicate which sets of observations it is advisable to repeat before including them with the others.

But if the graph is an absolutely unmistakable straight line its intersection $\bar{Q}$ with the axis of $\bar{X}$ can be read off the graph paper.

Having now obtained the scale-readings $\bar{P}$, $\bar{Q}$ we can at once calculate f from the relation $\frac{1}{f}=\frac{1}{u}+\frac{1}{v}=\frac{1}{\overline{PH_1}}+\frac{1}{\overline{QH_2}}$, by means of a four-figure table of reciprocals.

This is as good a value of f as you can get from a single position of the object. But it can be shown by the theory of probability that, so far as 'casual' experimental errors are concerned, the mean of n independent observations is likely to be $\sqrt{n}$ times as accurate as the result of a single observation. It is therefore worth while to repeat the foregoing procedure with the object pin at about $2f-1$ cm. from H_1, giving a value v_1 for the distance of H_2 from the focus conjugate to P; the value of f for these values of u and v should be calculated from the lens-formula. Then to set the object pin v_1 from H_1 and repeat the procedure; finally taking the mean of all three values of f as the focal length. It is shown theoretically in the appended note that in this way any error in the position of B has no effect on the result.

Degree of accuracy to be expected. Consider briefly the degree of accuracy which we are likely to attain. It is shown in the appended note that under these conditions errors in f arise only from errors in $\bar{P}$ and $\bar{Q}$ and that the error in f for one setting of P does not exceed one-quarter of the sum of those errors. The former is unlikely to be more than perhaps one-third of a millimetre; the latter should not exceed 2 mm.; hence the error in one determination of f should not exceed $\frac{2+\frac{1}{3}}{4}$ mm. If three independent measurements of f have been made, the accumulated errors should not exceed $\frac{1}{\sqrt{3}}$ of this, or $\frac{1}{3}$ mm., or an error of one in 300; if the work has been carefully done, it is most likely to be less.

Note on the Theory

(A) Consider a convergent lens whose faces are parts of two spheres of equal radius, made of glass whose index of refraction is μ. Denote its focal length by f, its thickness AB by t and the mid-point between A and B by C. Suppose that a ray passes through P_0, a point on the optic axis AB, and after passing through the lens cuts the optic axis again at Q, and that this emergent ray when produced backwards cuts the central plane through C at S, and that S is at a distance y from AB.

Denote by u_0 the distance of P_0 from a fixed point H_1, and by v the distance of Q from another fixed point H_2; it turns out in the course of the mathematical analysis that the relation between v and u_0 and y takes its simplest form if these so-called Principal Points H_1 and H_2 are fixed in such a position that

$$H_1C = CH_2 = \frac{\mu - 1}{2\mu} t, \tag{1}$$

and this will be assumed to be done.

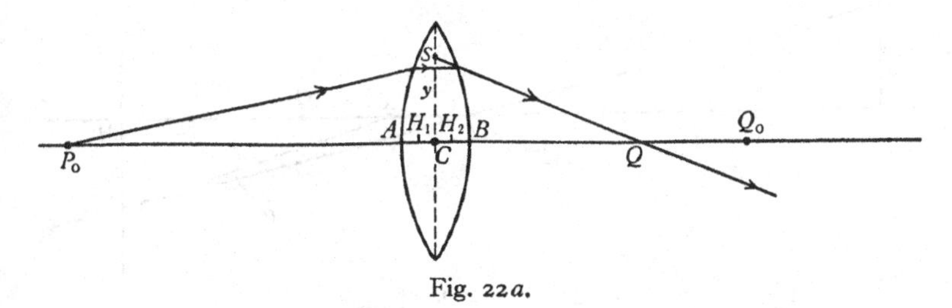

Fig. 22*a*.

This relation can be calculated, by means of Snell's Law and a great deal of simple geometry and algebra, in the form

$$\frac{1}{v} + \frac{1}{u_0} = A_0 + A_2\frac{y^2}{f^2} + A_4\frac{y^4}{f^4} + \text{higher powers of } \frac{y}{f}, \tag{2}$$

where A_0, A_2, etc., are constants, independent of v and y. If we put $y=0$ in (2) and denote the corresponding position of Q by Q_0 (so that Q_0 is the focus conjugate to P_0 for a thin co-axial pencil of rays), and the distance Q_0H_2 by v_0, then

$$\frac{1}{v_0} + \frac{1}{u_0} = A_0 = \frac{1}{f} \text{ on the ordinary definition of } f. \tag{3}$$

From (2) and (3) we can deduce a relation more suitable for our present purpose; writing γ for $\frac{v_0}{f}$,

$$\begin{aligned} QQ_0 = {} & \frac{1}{2f\mu}\left\{\frac{4\mu^3 - 4\mu^2 - \mu + 2}{4(\mu-1)^2}\gamma^2 - (3\mu+2)(\gamma - 1)\right\} y^2 \\ & - \frac{1}{f^2}\left\{1{\cdot}89\gamma^2 - 4{\cdot}98\gamma + 9{\cdot}76 - \frac{5{\cdot}62}{\gamma} + \frac{0{\cdot}35}{\gamma^2}\right\} ty^2 \\ & - \frac{1}{f^3}\left\{2{\cdot}78\gamma^3 - 8{\cdot}24\gamma^2 + 7{\cdot}90\gamma - 7{\cdot}75 + \frac{2{\cdot}38}{\gamma} - \frac{1{\cdot}60}{\gamma^2}\right\} y^4, \end{aligned} \tag{4}$$

where μ has been given its approximate value of $\frac{3}{2}$ in the small terms in ty^2 and y^4.

In this experiment we are mostly concerned with the particular case where $u_0 = v_0$, so that each equals $2f$; γ then becomes 2, and (4) reduces to

$$QQ_0 = \frac{\mu^2}{2f(\mu-1)^2}y^2 - \frac{11{\cdot}45}{f^2}ty^2 + \frac{1{\cdot}86}{f^3}y^4. \tag{5}$$

If $f = 10$ cm., $t = 0{\cdot}8$ cm., and $\mu = \frac{3}{2}$ and y is measured in centimetres this becomes

$$QQ_0 = 0{\cdot}448y^2 + 0{\cdot}0019y^4. \tag{6}$$

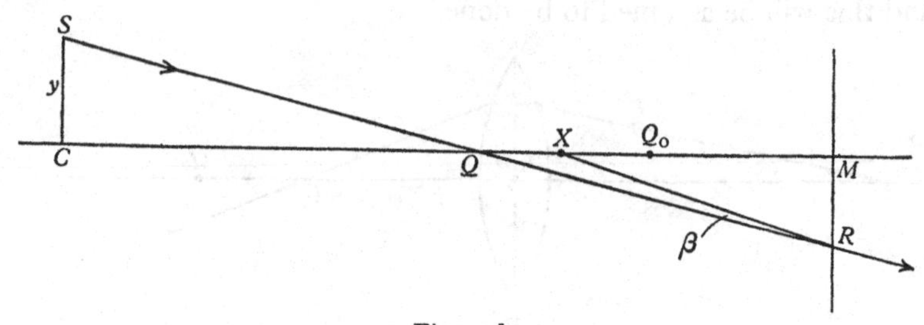

Fig. 22*b*.

Under these conditions the graph connecting QQ_0 and y^2 is almost exactly a straight line, though it is slightly curved owing to the presence of the small term in y^4. In the experiment we got a graph for values of y^2 ranging between 2·0 and 4·0, so that graph may be regarded as the tangent at $y^2 = 3{\cdot}0$ to the curve represented by (6). The error in taking the intersection of the axis of $\bar{X}$ and this tangent instead of the curve can easily be shown to be $0{\cdot}0019 \times 3{\cdot}0^2$ or 0·017 cm.; this is inappreciable with our equipment, and the 'extrapolation' in the experiment is fully justifiable.

It will be seen from (6) that this graph makes with the $\bar{X}$ axis an angle whose cotangent is about 0·45, or about 66°, so they make a 'good cut'.

(B) The method adopted in this experiment for picking out one ray SQ which started from P_0, and determining the corresponding pair of values of y and the scale-reading of Q is to put a search pin at an arbitrary point X in the optic axis CM in Fig. 22*b* and slide a pin-hole along a scale MR perpendicular to CM until when viewed through the pin-hole X seems to cover P_0; the ray SQ must then pass through X, and the corresponding value of y can be deduced from MR by similar triangles as shown in the instructions.

From a theoretical point of view this is quite satisfactory, but from a practical point of view we are almost equally concerned with the 'sharpness of setting' of the pin-hole in MR. Consider then the situation in Fig. 22*b*, before the apparent coincidence of P_0 and X is reached. By the geometry of Fig. 22*b*, by similar triangles

$$\frac{MR}{MQ}=\frac{y}{CQ}=\frac{y}{CX} \text{ nearly} \tag{7}$$

and $\beta = RXM - RQM = \frac{MR}{MX} - SQC = \frac{y\,MQ}{MX.CQ} - \frac{y}{CQ}$ nearly.

So $$\beta=\frac{y}{CQ}\,\frac{MQ-MX}{MX}=\frac{y\,QX}{CQ.MX}=\frac{y(Q_0X-QQ_0)}{CQ.MX}$$

$$=\frac{y(Q_0X-0{\cdot}45y^2)}{CX.MX} \text{ nearly.} \tag{8}$$

Hence, the rate of increase of the angular separation β with increase of MR

$$=\frac{d\beta}{dMR}=\frac{d\beta}{dy}\,\frac{dy}{dMR}=\frac{Q_0X-3\times 0{\cdot}45y^2}{CX.MX}\times\frac{CX}{MQ} \text{ by (8) and (7)}$$

$$=\frac{Q_0X-3\times QQ_0}{MX.MQ}.$$

Therefore, at coincidence, the rate of decrease of β with increase of MR, since QQ_0 then equals Q_0X, is $\frac{2\,Q_0X}{MX^2}$.

Hence the sharpness of setting of the pin-hole varies directly with the distance of X from Q_0, and inversely as the square of MX^2. Now the distance of X or Q from Q_0 varies directly with y^2, so we cannot expect the method to work at all well for small values of y^2; in practice, it turns out that the method is unsatisfactory if y^2 is less than about 2·0. And bringing M nearer to Q_0 in arranging the apparatus should markedly improve the sharpness of setting; but this also reduces the length MR, which must be read accurately; a compromise has therefore to be adopted, and a convenient one is given in the instructions.

(C) *Effect of errors of observation.* Suppose that for any particular setting of object pin and lens, our observations and calculations have given $\bar{P}$, $\bar{Q}$, $\bar{H}_1$, $\bar{H}_2$, B and f, and that these numbers are in fact too small by the small errors ΔP, ΔQ, ΔH_1, ΔH_2, ΔB and Δf, so that the true values are $\bar{P}+\Delta P$, etc.

Then the true u is $\bar{P}+\Delta P-(H_1+\Delta H_1)$ and the true v is

and $$H_2+\Delta H_2-(Q+\Delta Q),$$

$$\frac{1}{f+\Delta f}=\frac{1}{\bar{P}+\Delta P-(\bar{H}_1+\Delta H_1)}+\frac{1}{\bar{H}_2+\Delta H_2-(\bar{Q}+\Delta Q)}$$

$$=\frac{1}{\bar{P}-\bar{H}_1}-\frac{\Delta P-\Delta H_1}{(\bar{P}-\bar{H}_1)^2}+\frac{1}{\bar{H}_2-\bar{Q}}-\frac{\Delta H_2-\Delta Q}{(\bar{H}_2-\bar{Q})^2}\text{ nearly.}$$

Now $\Delta H_1=\Delta H_2=\Delta B$, since $\bar{H}_1$ and $\bar{H}_2$ were not observed but calculated from the measured value $\bar{B}$. Hence

$$\frac{1}{f+\Delta f}=\frac{1}{f}-\frac{\Delta f}{f^2}=\frac{1}{\bar{P}-\bar{H}_1}+\frac{1}{\bar{H}_2-\bar{Q}}-\frac{\Delta P}{(\bar{P}-\bar{H}_1)^2}$$

$$+\frac{\Delta Q}{(\bar{H}_2-\bar{Q})^2}+\Delta B\left\{\frac{1}{(\bar{P}-\bar{H}_1)^2}-\frac{1}{(\bar{H}_2-\bar{Q})^2}\right\}. \tag{9}$$

Our calculation of f was made from the equation

$$\frac{1}{f}=\frac{1}{\bar{P}-\bar{H}_1}+\frac{1}{\bar{H}_2-\bar{Q}}. \tag{10}$$

Hence, subtracting (9) from (10),

$$\frac{\Delta f}{f^2}=\frac{\Delta P}{(\bar{P}-\bar{H}_1)^2}-\frac{\Delta Q}{(\bar{H}_2-\bar{Q})^2}-\Delta B\left\{\frac{1}{(\bar{P}-\bar{H}_1)^2}-\frac{1}{(\bar{H}_2-\bar{Q})^2}\right\}.$$

The last term of this equation can be written

$$-\Delta B\left(\frac{1}{\bar{P}-\bar{H}_1}+\frac{1}{\bar{H}_2-\bar{Q}}\right)\left(\frac{1}{\bar{P}-\bar{H}_1}-\frac{1}{\bar{H}_2-\bar{Q}}\right)$$

or $$-\frac{\Delta B}{f}\left(\frac{1}{\bar{P}-\bar{H}_1}-\frac{1}{\bar{H}_2-\bar{Q}}\right)$$

by (10). Hence we get

$$\frac{\Delta f}{f^2}=\frac{\Delta P}{u_0^2}-\frac{\Delta Q}{v_0^2}-\frac{\Delta B}{f}\left(\frac{1}{u_0}-\frac{1}{v_0}\right)\text{ nearly.} \tag{11}$$

If, as in this experiment, we make $u_0=v_0=2f$, the last term vanishes and errors in measuring $\bar{B}$ have no effect on the result. We are left with the effect of errors in measuring $\bar{P}$ and $\bar{Q}$; the signs of these errors are of no importance since we have to consider only the worst case, in which an error does harm irrespective of sign.

Hence, under these conditions,

$$\Delta f=\frac{f^2}{4f^2}\Delta P+\frac{f^2}{4f^2}\Delta Q$$

$$\text{or error in } f=\tfrac{1}{4}(\Delta P+\Delta Q). \tag{12}$$

Again, (11) shows that we can eliminate the effect of small errors in locating B if we start with any value u_0 and derive the corresponding v_0, and then start with the object v_0 from H_1, and get (about) u_0 for the distance of the image from H_2. The mean of the two values of f thus obtained will be unaffected by ΔB. This device is adopted in the last stage of this experiment.

Exp. 21. *Newton's lens formula*

Set up a convergent lens of focal length (f) not exceeding 20 cm. at the middle of any form of optical bench. By means of a pin and plane mirror locate, as accurately as you can by the parallax method, its principal foci F_1 and F_2; record their scale-readings on the scale of the optical bench.

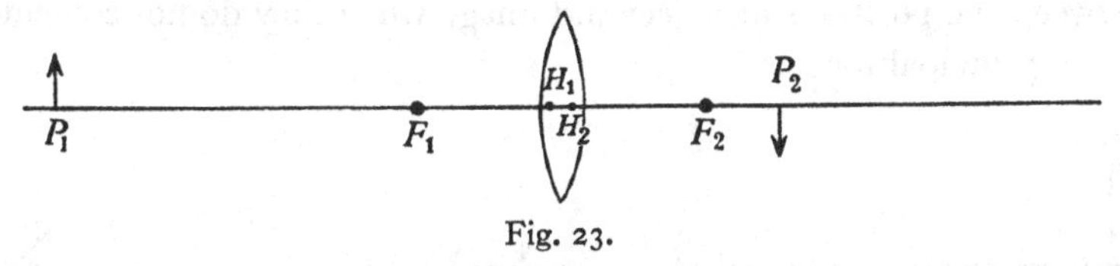

Fig. 23.

Set up a pin P_1 near one end of the bench; with another pin locate roughly its real image P_2; record the scale-reading of P_2; leaving the second pin P_2 undisturbed locate P_1 (now regarded as the image of P_2) accurately, and record its scale-reading.

Deduce the lengths P_1F_1 and P_2F_2 and calculate the value of $P_1F_1 \times P_2F_2$.

Repeat for a series of positions of P_1, until P_2 gets near the end of the scale; when P_2F_2 gets larger than P_1F_1, greater accuracy can be got by leaving P_1 undisturbed and adjusting P_2, since the parallax becomes more obvious when the image is larger.

It will probably be found that the product $P_1F_1 \times P_2F_2$ is approximately the same in each case; if so, calculate its arithmetic mean, and get the square root f of this mean from the tables.

This can be shown theoretically to be the focal length of the lens; if the lens is of negligible thickness, f will probably be found in this case to be approximately half of F_1F_2.

Hence, in general, $\mathbf{P_1F_1 \times P_2F_2 = f^2}$.

This is Newton's lens formula. It is in many respects superior to the better known formula $\frac{1}{u}+\frac{1}{v}=\frac{1}{f}$; it has the great practical advantage, when used for measuring f, that all measurements are made between pins, whose scale-readings can be determined easily and accurately, and no allowances have to be made either theoretically or practically for the thickness of the lens. This last point is established by the general theory given in Appendix B.

Exp. 22. *Lens and plane mirror*

One of the standard methods of measuring the focal length of a convergent lens is to use a plane mirror and thereby to get the real image of an object in front of the lens to coincide with the object. It is a very good method, if the mirror is a true plane and if the location of the image by the parallax method is skilfully carried out. It is in fact so good that it is worth while to find out more about it, and to recognize that it is a particular case of a more general relation between the positions of object and image when they do not coincide at the principal focus.

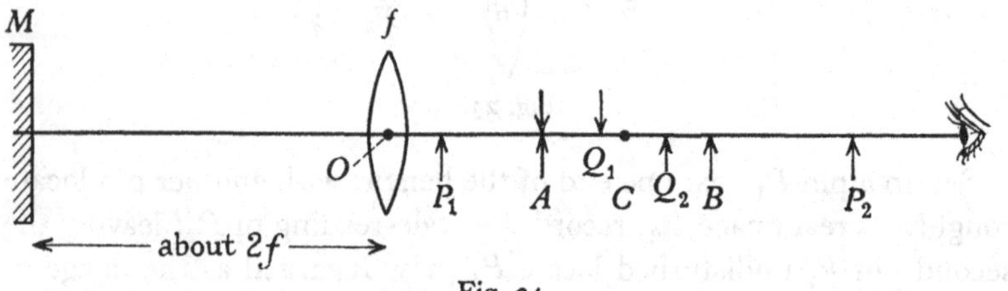

Fig. 24.

The following experiments show how this investigation can be conducted, and in particular how the general relation can be discovered by purely experimental means.

Take a convergent lens, whose focal length f can most conveniently be about 20 cm.; get a very rough value for f by focusing a distant object on a piece of paper. Set up the lens with its centre O about $2f$ from one end of the optical bench, and a plane mirror at a point M near the end of the bench, as in the diagram. Put an object pin at P_1, close to the lens; locate its image at Q_1 by means of a search pin. Note the scale-readings of P_1 and Q_1.

Repeat for a series of three or four positions of P_1 as it is moved away from the lens; Q_1 will be found to approach P_1. Get also the position A, where object and image coincide, by using the object pin as a search pin as in the ordinary process of finding the focal length of a convergent lens by means of a plane mirror.

Make a table of the scale-readings, including that of A, under the headings 'scale-reading of object' and 'scale-reading of image'; you should *not* tabulate the distances of P_1, Q_1 and A from O, or from any point other than the end of the ruler that forms the optical bench.

Next, move the object pin to P_2, about $3f$ from the lens, and locate the position of Q_2 the image of P_2. This image will probably be erect and rather small; in order to see it clearly it is advisable to hold a piece of white paper almost in the line between your eye and P_2, so that the image will be seen against a white background; the point of the object pin should be set at the level of the optic axis but a little to one side of it, and the point of the search pin should be a little below the optic axis; when the search pin is at Q_2 it will not then completely obscure the image of P_2 and any parallax between them can be seen as the eye is moved horizontally. In this part of the experiment great care in the use of the parallax method is necessary to obtain good results.

Note the scale-readings of P_2 and Q_2. Repeat for a series of three or four positions of P_2, as it is moved towards the lens. As before, Q_2 will be found to approach P_2. Get the position B where object and image coincide, by using the object pin as search pin.

Enter these readings on your table. Since object and image are interchangeable, you can double the number of entries in the table by merely copying into the 'image' column each entry under 'object', and vice versa.

Plot a graph connecting the corresponding scale-readings of object and image; you will probably get one like Fig. 25, in which $f=20{\cdot}0$ cm., the mirror was put at 1·0 cm. on the scale (and therefore not shown here), the lens was put at 37·0 cm. (shown at O on both axes) and A and B were found to be at 57·0 and 82·0 cm. respectively (as shown on both axes).

The aim of this experiment is to discover from a few observations in one particular case, and then to substantiate in general, an algebraical relation between the distance of an object from some fixed point on the optic axis of this set-up and the distance of the corresponding image from the same, or some other, fixed point. We have (1) to find the most suitable fixed point or points, and then (2) to discover the relation.

(1) The mirror M and the lens O would seem to be possible candidates. But neither shares fully in the striking general symmetry of the graph. The point C_1 midway between A and B does this, for the whole graph is symmetrical round it. Let us then adopt C in

Fig. 25 as the single base from which to measure, and see what happens. The next step is then to add two columns to the table of observations, by subtracting the scale-reading of C (the arithmetic mean of the scale-readings of A and B) from each of the first two columns; let us denote the results by p and q, the respective distances of object and image from C.

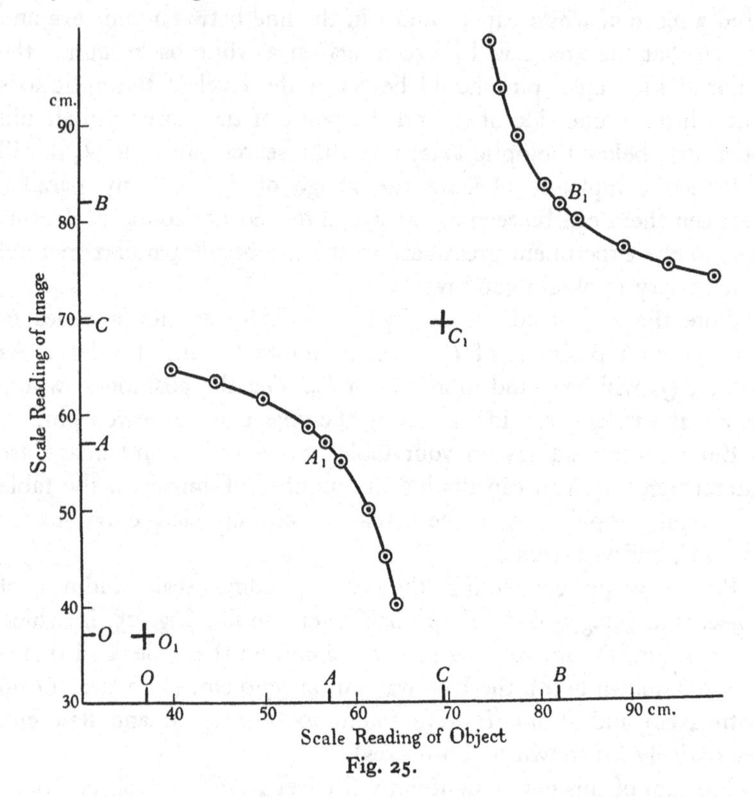

Fig. 25.

(2) A cursory inspection of these two new columns shows that as p decreases q increases; a little very rough mental arithmetic suggests that the product pq may be constant. Calculate this product accurately in each case (for one-half of the table, since the other half is a mere copy of it) making a fifth column, headed pq.

It is worth while to determine the degree of accuracy of your observations, on the assumption that the product pq should be constant. Find the arithmetic mean of the numbers in the last column (which will all be positive), and make a sixth column showing (without regard to sign) the differences between this and the numbers in the fifth column. The arithmetic mean of the numbers in this sixth column will be the Mean Error of your experiments, and from it and your mean value of pq you can easily find the mean percentage error (see Introduction, § 1 (*b*)).

But a more important question is whether the differences in the sixth column are random, or seem to change according to the changes in the first column. If the former is the case, they probably arise from inaccuracies in the experiment; if the latter, the assumption that pq = constant may be unfounded.

If it turns out that you are justified in assuming that pq is constant in the particular set-up with which you started, you are justified, since that set-up was chosen at random, in asserting that the law should hold good for all similar set-ups; the value of the constant pq may well, of course, depend on the details.

The appended note on the theory shows how it does so; and it shows, by rather complicated reasoning, that if the object is nearer to the lens than C its image should be inverted, while if it is on the other side of C its image should be erect. Your experiments will doubtless have led you, much more simply, to the same conclusion.

Assuming that pq = constant, if the object is at C, so that $p=0$, q becomes infinitely large. Hence C is the principal focus of the combination, as ordinarily defined.

Note on the Theory

Let P denote an erect object u in front of the lens and let P' be its lens-image, u' behind the lens. Let Q' (v' behind the lens) be the image of P' in the mirror M, situated a behind the lens, and let Q (v in front of the lens) be the image of Q' formed by the lens.

Then by the ordinary lens formula

$$\frac{1}{u'}+\frac{1}{u}=\frac{1}{f}, \quad \text{or} \quad u'=\frac{uf}{u-f},$$

and similarly

$$v'=\frac{vf}{v-f}.$$

Since Q' is the mirror image of P', $v'-a=a-u'$.

Hence

$$2a=u'+v'=\frac{uf}{u-f}+\frac{vf}{v-f},$$

or

$$2a\{uv-(u+v)f+f^2\}=f\{uv-uf+uv-vf\}=f\{2uv-(u+v)f\},$$

or

$$uv2(a-f)-(u+v)(2a-f)f=-2af^2,$$

or
$$uv-(u+v)\frac{(2a-f)f}{2(a-f)}=-\frac{af^2}{a-f},$$

or
$$\left\{u-\frac{(2a-f)f}{2(a-f)}\right\}\left\{v-\frac{(2a-f)f}{2(a-f)}\right\}=\frac{(2a-f)^2f^2}{4(a-f)^2}-\frac{af^2}{a-f}=\frac{f^4}{4(a-f)^2}. \quad (1)$$

Parts of the curve for which this is the equation are sketched in Fig. 25.

Putting $u=v$, which corresponds to making object and image coincide, we get, by taking the square root,

$$\left.\begin{aligned} u-\frac{(2a-f)f}{2(a-f)}&=\pm\frac{f^2}{2(a-f)}.\\ \text{Hence}\quad u=f\quad&\text{or}\quad\frac{af}{a-f},\end{aligned}\right\} \quad (2)$$

corresponding to the two points A and B in Fig. 24.

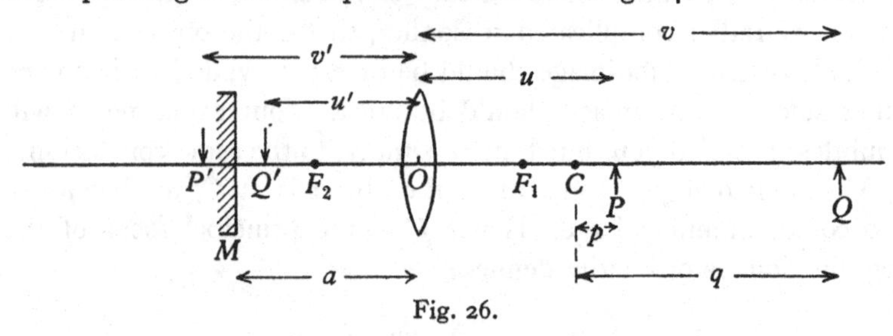

Fig. 26.

The former value of u corresponds to that found in the ordinary method of measuring f by a plane mirror; it does not involve a, so the mirror can be put at any distance behind the lens in using that method.

When $u=\frac{af}{a-f}$, $\frac{1}{f}-\frac{1}{a}=\frac{1}{u}$ or $\frac{1}{u}+\frac{1}{a}=\frac{1}{f}$, so that the first image P' is at a distance a behind the lens and therefore lies on the reflecting surface of the mirror; this image acts as an object whose lens-image coincides with the original object.

It will be seen from (2) that unless a is greater than f, B will be behind the lens and the second branch of the graph will refer to virtual images, which are outside the scope of this experiment. So the mirror must be put to the left of F_2 in Fig. 26.

Equation (1) will be simplified if we put

$$p=u-\frac{(2a-f)f}{2(a-f)}\quad\text{and}\quad q=v-\frac{(2a-f)f}{2(a-f)},$$

which is equivalent to transferring the origin to a point (C_1 say) whose co-ordinates with respect to O are $\frac{(2a-f)f}{2(a-f)}, \frac{(2a-f)f}{2(a-f)}$.

Equation (1) then becomes

$$pq=\frac{f^4}{4(a-f)^2}. \tag{3}$$

This is the equation of a rectangular hyperbola, with the axes as asymptotes.

This transfer of origin means that, in Fig. 26, we now measure the distances of object and image (p and q) from a new point C at a distance $\frac{(2a-f)f}{2(a-f)}$ from O. Since

$$CF_1=OC-OF_1=\frac{(2a-f)f}{2(a-f)}-f=\frac{f^2}{2(a-f)},$$

we can write (3) in the form

$$pq=CF_1^2. \tag{4}$$

It will be seen from (2) that C is the mid-point between A and B.

It will also be seen from (1) that since the right-hand side is a square, and therefore necessarily positive, u and v must be both larger or both smaller than $\frac{(2a-f)f}{2(a-f)}$, or OC. *Hence P and Q must lie on the same side of C.* (5)

Nature of images

Upper branch. Suppose that the object P is farther from the lens than C. Then p is positive and therefore q is positive, and we are dealing with points on the upper branch of the curve. From (5) Q also is farther from the lens than C; hence object and final image are both outside of the principal focus; hence their lens-images are both erect. These lens-images are the mirror-images of one another, so they must be the same way up; hence, the object and final image are the same way up. Hence, throughout the upper branch the image is *erect*.

Lower branch. Writing (4) in the form $p/CF_1=CF_1/q$ we see that if p is numerically greater than CF_1, q must be numerically less than CF_1. Hence, if the object lies between the lens and its principal focus, as P_1 in Fig. 24 (so that p and q are both negative and we are dealing with the lower branch), the final image will lie between C and the principal focus, as Q_1 in Fig. 24. So the lens-image of the object will be virtual and erect, while the lens-image of the final image will be inverted. As before, mirror-images are the same way up, so the object and final image are opposite ways up. Since object and final image are interchangeable, this holds good also if it is the object that lies between C and the principal focus. Hence, throughout the whole of the lower branch the image is *inverted*.

Exp. 23. *Focal length of a convex mirror*

The following experiment illustrates the procedure necessary to get the most accurate results out of simple apparatus used in a standard method. It may be felt that the increase in accuracy demands a disproportionate amount of care, but this is usually the case.

Set up on an optical bench a convergent lens, of focal length preferably less than 15 cm. Find by a plane mirror and the parallax method the scale-readings of its two principal foci F_1 and F_2; since the accuracy of the whole of this experiment depends largely on the accuracy with which these foci are located it is worth while to use the 'indirect parallax method' of Exp. 20. Denote by f_1 the focal length of the lens, deduced from the distance between F_1 and F_2 and the thickness of the lens; it is shown in the appended note that theoretically $f_1 = \frac{F_1F_2}{2} - \frac{\text{thickness of lens}}{6}$.

Remove the plane mirror and pins, and set up the convex mirror, whose focal length (f_2) is to be found, facing the lens with its reflecting surface passing through F_2 (see Fig. 27).

Put an object pin at P in front of the lens, farther from it than F_1, take the scale-reading of the pin and deduce the distance (x) of the object from F_1. By means of the direct parallax method, adjust a search pin to coincide with the real image of the object P formed by two passages of the light through the lens and one reflection at the mirror; deduce the distance (y) of this image from F_1.

Repeat for five or six values of x, and in each case calculate $x+y$.

(A) If $x+y$ has a nearly constant value. The appended note on theory shows that $x+y$ should be constant, and that $f_2 = \frac{f_1^2}{x+y}$. Calculate the mean value of $x+y$. Hence, and from the measured value of f_1, calculate f_2.

(B) If $x+y$ is not nearly constant. Plot a smoothed graph connecting x and y. If this best-fit line is straight, the value of f_2 found in (A) is about as accurate as can be deduced from your experiments, though it is possible that the best-fit line found by the method of Introduction (5) will give a more accurate value for $x+y$, as the mean of the intercepts on the axes.

But if the best-fit line is curved, the most accurate value for $x+y$ will be twice the ordinate of the point of intersection of the curve and the line whose equation is $x=y$.

With reasonable care the variations in $x+y$ will be so small that (A) should give a very satisfactory result.

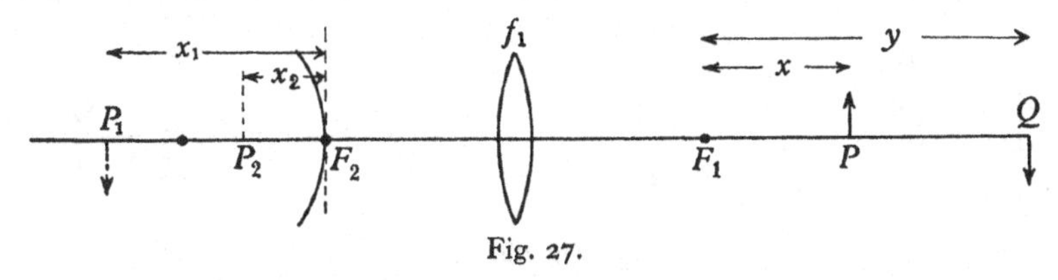

Fig. 27.

Note on the Theory

(i) It is shown in Appendix B (i) that, if F_1 and F_2 are the two principal foci of a convergent lens and H_1 and H_2 are its principal points, the distance $F_1F_2=2f_1+H_1H_2$ and H_1H_2 is approximately one-third of the thickness of the lens. Hence

$$f_1=\frac{F_1F_2}{2}-\frac{\text{thickness of lens}}{6}.$$

(ii) Suppose that the object is at P, x from F_1, and that its image formed by one passage of the light through the lens is at P_1, distant x_1 from F_2, then by Newton's lens formula (see Exp. 21)

$$x\,x_1=f_1^2. \qquad (1)$$

This image at P_1 is the object for the reflection in the mirror, of numerical focal length f_2; hence, if the image formed by the mirror is at P_2 distant x_2 from the pole F_2 we have

$$\frac{1}{x_1}+\frac{1}{x_2}=\frac{1}{f_2}. \qquad (2)$$

This image at P_2 is the object for the lens; suppose that its image is at Q at a distance y from F_1, then by Newton's formula

$$x_2y=f_1^2. \qquad (3)$$

From (1) $\quad \frac{1}{x_1}=\frac{x}{f_1^2}, \quad$ and by (3) $\quad \frac{1}{x_2}=\frac{y}{f_1^2}.$

Hence by (2) $\quad \frac{x}{f_1^2}+\frac{y}{f_1^2}=\frac{1}{f_2} \quad$ or $\quad x+y=\frac{f_1^2}{f_2} \quad$ or $\quad \mathbf{f_2=\frac{f_1^2}{x+y}}.$

Exp. 24. *Lens formula for a concave lens*

Take a plane mirror and a convergent lens A of 15 to 25 cm. focal length, and set them up as in the diagram, omitting B. By means of a pin locate, by the parallax method, F_1 the principal focus of A; note the precise scale-reading of F_1 and remove the pin. You have found the point from which rays must diverge if they are to converge to the same point after their two passages through A and one reflection at the mirror.

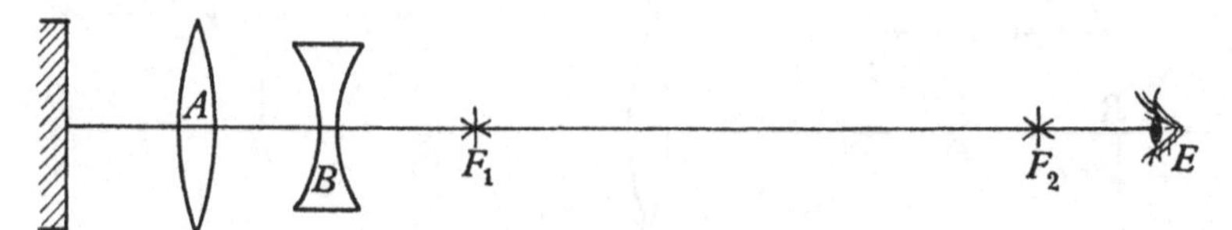

Fig. 28.

Take a divergent lens, whose focal length may conveniently be from 10 to 20 cm.; it should be less than that of the convergent lens; a good combination is 20 cm. convergent and 15 cm. divergent. Set it up with its centre at B, between A and F_1, and near to A. Set up a pin at F_2 near the end of the bench; increase AB until an eye at E sees a real image of this pin coincident with the pin; make the final adjustment for coincidence by moving the pin at F_2. Note the precise readings of B and F_2.

Repeat for a series of about nine positions of B, until B comes to about 6 cm. from F_1.

With the help of a table of reciprocals make a table of corresponding values of BF_1, BF_2, $\frac{1}{BF_1}$ and $\frac{1}{BF_2}$.

Plot the points whose co-ordinates are $\frac{1}{BF_1}, \frac{1}{BF_2}$.

You will probably find that these points lie approximately on a straight line which is equally inclined to the axes; hence, if this line intersects the axis of $\frac{1}{BF_1}$ at a distance which we may denote by $-\frac{1}{f_2}$, the relationship is $\frac{1}{BF_1} - \frac{1}{BF_2} = -\frac{1}{f_2}$, where f_2 is a constant.

If then we can assume that the relation which we have found for a limited range of positions of B for a particular concave lens holds good for all such positions and lenses, which is a reasonable assumption since we took these conditions at random, we have found a relation applicable in all cases, for this set-up of two lenses and a mirror.

In the first part of the experiment we found that when a divergent pencil which fell on A from the right, after passing twice through A and being reflected once by the mirror, retraced its original course and converged again on the same point from which it started, it could only do so if that point was F_1. This must hold good even if the divergent pencil did not actually start from F_1, but was produced by a lens B, and if the convergent pencil did not later actually reach F_1 but was intercepted by the lens B.

In the second part of the experiment a divergent pencil produced between B and A by the lens B, after passing twice through the lens A as above, must have retraced its original course between A and B, since it retraced its course between A and F_1, and rays of light are reversible. Consequently in this part of the experiment a pencil diverging from F_2 must have diverged from F_1 after passing through the concave lens B, so that F_1 must be the virtual focus conjugate to F_2 for that concave lens; hence, the above relation between BF_1, BF_2 and f_2 holds good for the conjugate foci F_1 and F_2 of the lens B. The mirror and lens A were merely auxiliaries to prove this.

Further, if BF_2 is infinitely large, so that $\frac{1}{BF_2}=0$, BF_1 is numerically equal to the constant f_2, and f_2 has the meaning ordinarily attached to the focal length of a lens, as the distance from the lens of the point to which a parallel axial pencil will converge (in the case of a convex lens), or of the point from which it will diverge, after passing through a concave lens.

It will be observed that this experiment provides a method of establishing the ordinary lens formula for a concave lens, without using any mathematical deduction from the laws of refraction of light.

Exp. 25. *Focal length of a concave lens*

The measurement, to a reasonable degree of accuracy, of the focal length of a concave lens presents greater difficulties than the same operation on a convex lens. Standard text-books give several methods, and if they are applied in succession to the same lens there is likely to be a depressing variety in the results.

Almost any method involves the use of a convex lens or concave mirror as an auxiliary; in many methods the focal length of this auxiliary has to be used in calculating the final result, so that any error in it increases the errors in the focal length of the concave lens. Several methods also involve a plane mirror, and any small curvature of that mirror is liable to introduce a further error into the result. So it is important to select the method that avoids such sources of error; there are two which theory and experience show to be capable of producing the most concordant results.

(A) The first of these is to form a combined lens by putting in contact with the concave lens, whose focal length f_2 we seek, a stronger convex lens, whose focal length f_1 we have measured. This forms a convergent lens whose focal length F we can measure. If we know the relation between F, f_1 and f_2 for this combined lens we can then deduce f_2.

It is commonly assumed that this relation is $1/F = 1/f_1 + 1/f_2$, but it is shown in Appendix B (ii) that if the thickness of the combined lens is t, it is more accurate to take the relation as being

$$\frac{1}{f_2}\left(1-\frac{t}{3f_1}\right)=\frac{1}{F}-\frac{1}{f_1}.$$

f_1 should be measured as in Exp. 20 by setting up a pin about $2f_1$ behind the convex lens, locating its image by the 'indirect parallax method', and using the ordinary lens formula; allowance must be made for the thickness of this lens by measuring u and v from its nearer faces and adding one-third of its thickness to u and to v before substituting in the formula.

The indirect parallax method can also, and should, be used in measuring F; a less accurate method of measuring F is that of setting up two pins d cm. apart, d being rather more than $4F$, and finding the two positions, l cm. apart, of the lens which make these two pins become conjugate foci, so that either is at the real image of the other. It is shown in the attached note that F is then

$$\frac{(d-\frac{1}{3}t)^2-l^2}{4(d-\frac{1}{3}t)}.$$

The main inconvenience of this method is that it needs an unusually long optical bench; for example, if $f_1 = 10$ cm. and $f_2 = -15$ cm., and $t = 0{\cdot}9$ cm., F will be $29{\cdot}1$ cm. and d will be about 120 cm.

(B) The second method follows the lines of Exp. 24. The scale-reading of F_1 in that experiment should be found by locating F_1 by the indirect parallax method if possible; if not, by the direct parallax method. There is no need to calculate f_1; if the mirror is only nominally plane and actually has some curvature this will combine with f_1 to form a convergent system of unknown power, which, however, has its principal focus at F_1; the scale-reading of that point is all we need to know about this convergent system. The only substantial error will come from uncertainty in locating F_2 and may be reduced by intelligent use of the parallax method.

In order to reduce the effect of errors as much as possible both BF_1 and BF_2 (they increase together) should be large, so that it is advisable to use a convex lens of focal length longer than f_2; this will give freedom of movement to the concave lens.

Take a concave lens, of nominal focal length 10 or 15 cm., and measure its focal length by combining it with a more powerful convex lens, as in (A); to save time, use a convex lens whose focal length has already been accurately measured by the method of Exp. 20. Measure also its focal length by method (B) and compare the two values.

Note on the Theory

Let H_1, H_2 denote the positions of the two principal points (see Appendix B (i)) of the lens in one of the positions for which the two fixed points P_1, P_2 are its conjugate foci, and let H_1', H_2' denote the positions of the same principal points in the other position of the lens for which P_1, P_2 are its conjugate foci.

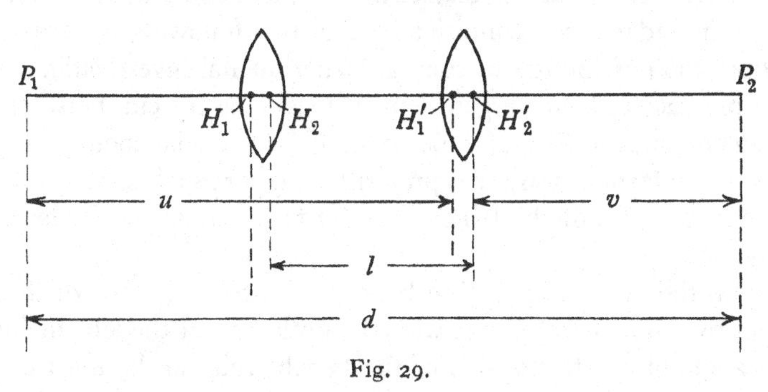

Fig. 29.

Then by symmetry, $P_1H_1' = P_2H_2 = u$, say. Denote P_2H_2' by v.

Denote P_1P_2 by d, and the distance through which any point on the lens, such as H_1, must be moved to pass from one position to the other, by l. $H_1'H_2'$ is one-third of the thickness (t) of the lens, so $H_1H_2 = \frac{1}{3}t$.

Hence

$$d=P_1H_1'+H_1'H_2'+P_2H_2'=u+\tfrac{1}{3}t+v \quad \text{or} \quad \mathbf{u+v=d-\tfrac{1}{3}t}.$$

And

$$l=H_2H_2'=P_2H_2-P_2H_2'=P_1H_1'-P_2H_2'=u-v \quad \text{or} \quad \mathbf{u-v=l}.$$

Hence, adding, $2u=d-\frac{1}{3}t+l$, and subtracting, $2v=d-\frac{1}{3}t-l$ and therefore multiplying, $4uv=(d-\frac{1}{3}t)^2-l^2$.

Now if F is the focal length of the lens $\frac{1}{F}=\frac{1}{u}+\frac{1}{v}$, so $F=\frac{uv}{u+v}$.

Therefore

$$F=\frac{(d-\frac{1}{3}t)^2-l^2}{4(d-\frac{1}{3}t)}.$$

Exp. 26. *Oblique incidence on a lens*

The elementary theoretical treatment of the behaviour of lenses is for the most part restricted to their effect on beams of light along, or inclined at a very small angle to, the optic axes of the lenses. This is a perfectly reasonable course of action since the mathematics involved in a more comprehensive treatment is liable to become intolerably ponderous. But in the practical use of lenses there is no such restriction on the beams of light with which they have to deal; since the theoretical approach to the consequent problems is barred by tacit consent, it is worth while to investigate them practically, so far as can be done, and to draw whatever general conclusions we can from these experiments that will enable us to understand and predict their behaviour under given conditions. The two following experiments illustrate the possibilities of such an experimental investigation.

On one face of a convergent lens of focal length f cm. between 15 and 20 cm. stick a piece of paper with a circular hole about 1 cm. in diameter, to form a stop; or if preferred, support such a stop with its centre in the axis of the bench, and its face parallel to the lens, as in Fig. 30.

Set up the lens on an optical bench as in the diagram, which is a section by a horizontal plane, with the optic axis of the lens inclined at an angle of 30° to the axis of the bench; this can be most easily done with a 60° set square, or a piece of card cut to this angle and laid along the edge of the block carrying the lens.

Find the scale-readings of A and B, the intersections of the axis of the bench with the faces of the lens; this can best be done with a knitting-needle of measured length held along the axis of the bench, with one end touching A and B in turn, the scale-reading of the other end being found with a set square with one edge along the scale of the bench.

(A) At P_0, about $2f$ from A, set up a cardboard or wooden screen with a square hole, each side about 1 cm., through it and a piece of fine wire gauze over the hole, the wires and edges of the hole being vertical and horizontal.

Measure P_0A and denote its numerical value by u cm. Put a lamp L (say 25 watt, pearl) at L behind P_0, and a white screen at Q; the room should, if possible, be dimly lit.

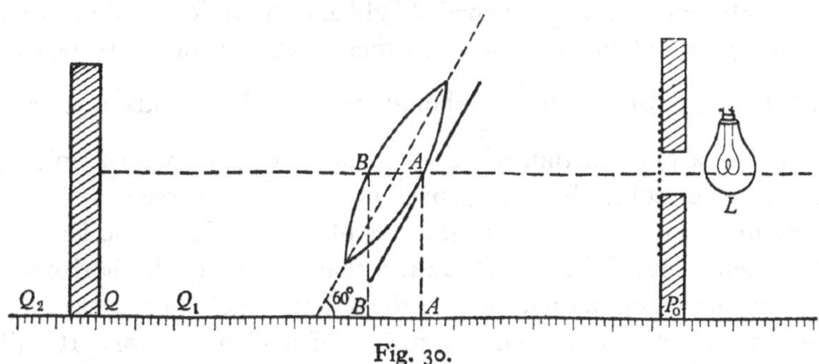

Fig. 30.

Move the screen Q up and down the bench and watch the changes in its illumination. It will be seen that there is no position of Q at which the hole and gauze give an image in sharp focus in all directions at once; but at some point Q_1 the image of the hole will be drawn out into a rectangle with its longer side vertical; the vertical edges will then be sharp and the horizontal ones blurred.

Viewing the image on the screen through a short (say 5 cm.) focus lens, adjust Q_1 so that the vertical wires in the image of the wire gauze are in the best focus, while the horizontal wires are completely out of focus. Q_1 is then in one sense an image of P_0; measure BQ_1 and denote its numerical value by v_1 cm.

Find from reciprocal tables the value of $\frac{1}{u}+\frac{1}{v_1}$, denote it by $\frac{1}{f_1}$ and find f_1 from the tables.

Repeat for two or more values of u. It is probable that f_1 will have about the same value in each case; if so, calculate their arithmetic mean.

At some other point Q_2 the image of P_0 will be drawn out horizontally; adjust Q_2 to get the horizontal wires in the best focus, and determine the corresponding value of f_2, as before.

These experiments are analogous to the experimental verification that f in the relation $\frac{1}{f}=\frac{1}{u}+\frac{1}{v}$ is constant when the incidence of the light on the lens is direct, and to the method of measuring the focal length f. But in the case of oblique incidence we find that there are two focal lengths; it is usual to call f_1, the smaller, the 'primary', and f_2 the 'secondary', focal length.

Calculate the ratio $\frac{f_1}{f_2}$ and compare it with $\cos^2\phi$, where ϕ is the angle between the incident pencil of light and the optic axis of the lens.

The appended note indicates that there should theoretically be two such focal lengths and that $\frac{f_1}{f_2}$ should theoretically be equal to $\cos^2\phi$.

(B) Pin a piece of thin white cardboard, or paper, with a hole in its centre, to cover the wire gauze and serve as a screen; set up a plane mirror perpendicular to the axis of the bench, on the far side of the lens from P_0; move P_0 until a sharp image of the horizontal wires is formed on the screen, as in the ordinary method of measuring the focal length of the lens by means of a plane mirror. It will probably be found that P_0 is at a distance of f_2 from A.

Repeat this for the vertical wires, and it will probably be found that P_0 is then at a distance of f_1 from A.

There is nothing in the preceding experiment (A) nor in the appended note on theory to show that the focal lines (see that note), which they show to exist on the left-hand side of the lens in Fig. 30, will, after reflection in the plane mirror, give rise to corresponding focal lines on the right-hand side of the lens. This experiment (B) shows that they do, and that there is the same relation between their distances from the lens. As in the case of direct incidence, P_0 was brought to the principal focus of the horizontal wires in the first case, causing the image of those wires to go to infinity, and the mirror reversed the rays, for the lens to bring them to a focus (sharp horizontally) at P_0.

As a matter of fact the full theory predicts this, though it is not mentioned in the note.

These experiments show the difficulties to be faced in designing a wide-angle photographic lens, which has to give a sharp image up to the edges of the plate; the image there is formed by rays which pass through the lens obliquely, and under these conditions a single lens does not give an image which is sharp in all directions.

Note on the Theory

The general theory on which this experiment depends involves mathematical considerations of such difficulty as to be unsuitable for this book. But some of the steps are elementary, and the results of the experiment may perhaps be more intelligible if they are set out here.

Let us confine our attention to a thin pencil of rays diverging from a single point P_0, with its central ray P_0A striking the surface of a thin convergent lens at A, passing through the centre C of the lens and emerging from it at B along the line BQ. Then if the angle of incidence of PA on the first face (whose radius is r) is ϕ, the elementary theory shows that BQ is parallel to P_0A, with an angle of emergence ϕ from the second face (whose radius is s).

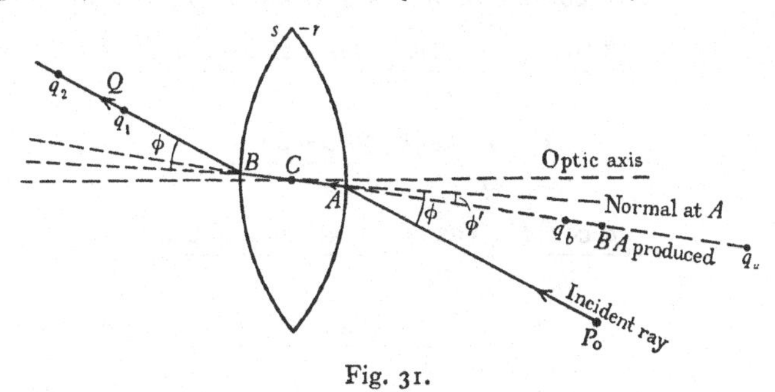

Fig. 31.

We will adopt the sign convention that distances are positive in the direction from which light is coming.

Denote by ϕ' the inclination of AB, the ray within the lens, to the normals at A or B; then $\sin\phi=\mu\sin\phi'$.

It can be shown that if the pencil of rays from P_0 covers a small square patch of the face of the lens round A, with one of its sides in the plane of the paper in the above diagram, every ray in the pencil, while it is within the glass, if produced backwards passes through two short straight lines, at right angles to one another, passing through q_a and q_b respectively; q_a and q_b are on the central ray AB produced backwards. These are called 'focal lines'; denote the distances from A of P_0, q_a and q_b by u, v_a and v_b respectively.

It can be shown that

$$\frac{\mu\cos^2\phi'}{v_a}-\frac{\cos^2\phi}{u}=\frac{\mu\cos\phi'-\cos\phi}{r} \qquad (1)$$

and

$$\frac{\mu}{v_b}-\frac{1}{u}=\frac{\mu\cos\phi'-\cos\phi}{r}. \qquad (2)$$

It may be observed in passing that if $\phi=0$, and so $\phi'=0$ and $\cos\phi=\cos\phi'=1$, then $v_a=v_b$ and we get the ordinary relation for direct incidence

$$\frac{\mu}{v}-\frac{1}{u}=\frac{\mu-1}{r}.$$

The lines through q_a and q_b are usually called the 'primary' and 'secondary' focal line respectively. It can be shown (and this is a more difficult step) that during the passage of the pencil through the second face of the lens, the primary focal line gives rise to a corresponding primary focal line, situated at q_1 on the emergent ray BQ, at a distance v_1 from B; similarly, the secondary focal line gives rise to a secondary focal line v_2 from B. These obey the same laws as in (1) and (2) (but, of course, with $1/\mu$ in place of μ), so that

$$\frac{1/\mu\cos^2\phi}{v_1}-\frac{\cos^2\phi'}{v_a}=\frac{1/\mu\cos\phi-\cos\phi'}{s}$$

and

$$\frac{1/\mu}{v_2}-\frac{1}{v_b}=\frac{1/\mu\cos\phi-\cos\phi'}{s},$$

or

$$-\frac{\mu\cos^2\phi'}{v_a}+\frac{\cos^2\phi}{v_1}=-\frac{\mu\cos\phi'-\cos\phi}{s} \tag{3}$$

and

$$-\frac{\mu}{v_b}+\frac{1}{v_2}=-\frac{\mu\cos\phi'-\cos\phi}{s}. \tag{4}$$

Adding (1) and (3) we get

$$\frac{\cos^2\phi}{v_1}-\frac{\cos^2\phi}{u}=(\mu\cos\phi'-\cos\phi)\left(\frac{1}{r}-\frac{1}{s}\right) \tag{5}$$

and

$$\frac{1}{v_2}-\frac{1}{u}=(\mu\cos\phi'-\cos\phi)\left(\frac{1}{r}-\frac{1}{s}\right). \tag{6}$$

If $\phi=\phi'=0$, we get $v_1=v_2$ and $\frac{1}{v}-\frac{1}{u}=(\mu-1)\left(\frac{1}{r}-\frac{1}{s}\right)=\frac{1}{f}$, as usual, if f is the focal length of the lens for direct incidence. So we can write (5) and (6) in the form

$$\frac{1}{v_1}-\frac{1}{u}=\frac{\mu\cos\phi'-\cos\phi}{(\mu-1)\cos^2\phi}\frac{1}{f} \tag{7}$$

and

$$\frac{1}{v_2}-\frac{1}{u}=\frac{\mu\cos\phi'-\cos\phi}{\mu-1}\frac{1}{f}. \tag{8}$$

The expressions on the right-hand sides of (7) and (8) are constants if ϕ is constant; it is convenient to denote them by $\frac{1}{f_1}$ and

$\frac{1}{f_2}$, treating f_1 and f_2 as the primary and secondary focal lengths of the lens for an obliquity ϕ of the central ray. Then (7) and (8) become

$$\frac{1}{v_1}-\frac{1}{u}=\frac{1}{f_1} \quad \text{and} \quad \frac{1}{v_2}-\frac{1}{u}=\frac{1}{f_2}.$$

It will be seen that $\frac{f_1}{f_2}=\cos^2\phi$, and that $f_1=\frac{(\mu-1)\cos^2\phi}{\mu\cos\phi'-\cos\phi}\times f$ and $f_2=\frac{\mu-1}{\mu\cos\phi'-\cos\phi}\times f$. Both f_1 and f_2 are less than f; f_1 is the smaller, but $f_1=f_2=f$ when $\phi=0$.

Exp. 27. *Focal lines*

Set up a convergent lens, of focal length (f) from 15 to 20 cm., with its optic axis inclined at 30° to the axis of an optical bench, as in Exp. 26. Find the scale-readings of A and B as in that experiment.

At about $2f$, from the lens and at right angles to the axis of the bench, set up a piece of paper carrying some large-type printing; put the lines of print horizontal and upside down. Look at this printing through the lens, with one eye, from a distance of about $3f$ from the lens.

The image of the print will probably appear to be sharp in every direction. Yet we know from Exp. 26 that this lens forms no sharp image of the print, though horizontal and vertical lines are sharply focused at two different points along the axis.

This apparent contradiction calls for explanation, which can be based on the following experiments.

Rule with a pencil on a piece of paper a cross with arms HH_1 and VV_1 at right angles to one another, HH_1 and VV_1 being each about $1\frac{1}{2}$ in. long and $\frac{1}{8}$ in. wide, and ink in the area within the boundaries of the cross, keeping the edges sharp and straight; set this up in place of the printed paper, with HH_1 horizontal. Look at this through the lens with one eye as before and note whether the image appears sharp throughout.

Exp. 26 shows that the lens forms a sharp image of the vertical arm VV_1 at Q_1 (see Fig. 30); if we set up a vertical pin with its point in the axis of the bench, between the eye and the lens, we should be able to locate Q_1 by moving the pin along the bench until there is no parallax between pin and image as the eye moves horizontally from side to side. Take the scale-reading of Q_1 and calculate

f_1 as in Exp. 26; it will probably be found to be approximately the same as was obtained in Exp. 26 if the same f and ϕ are used.

When this pin has thus been fixed at Q_1, if the eye is moved vertically up and down, it will be seen that there is a great deal of parallax between the point of the pin and the image of the horizontal arm HH_1; hence this pin does not coincide with the sharp image of HH_1.

Leaving this pin undisturbed, set up another vertical pin, with its point at the same level as the centre of the lens but slightly to one side of the axis of the bench; move it until its point shows no parallax with the image of HH_1 as the eye is moved vertically up and down. Take the scale-reading of this pin and calculate the value of f_2, as in Exp. 26, and compare it with f_2 as found in that experiment.

You have now got two pins, one situated at the sharp image of each arm of the cross; if the eye is moved up and down, one pin shows no parallax with the image of the cross, while the other moves rapidly relatively to that image; if the eye is moved sideways, the pins exchange their rôles. Hence, although to the unaided eye there appears to be only one image of the cross, we find that there are in fact two, some distance apart, with different properties in the two directions, vertically and horizontally.

This is consistent with the theory stated in the note appended to Exp. 26 (though the note does not contain a proof of the fact) that all the rays diverging from a single point on the edge of one of the arms of the object, after passing through the lens pass through a short vertical 'focal line' at Q_1 and later through a short horizontal 'focal line' at Q_2. If we consider a succession of such points on the edge of the vertical arm VV_1 the focal lines at Q_1 will all lie along one single line, and they will combine to form a sharp image of VV_1 at Q_1, while the focal lines at Q_2 will be parallel to one another and will not form a sharp image of VV_1 there; its edge will be very blurred.

Similarly, if the points lie along the edge of HH_1, this theory of focal lines calls for a blurred image of HH_1 at Q_1 and a sharp image at Q_2.

Now remove the cross from the bench, to prevent distraction, and look at the two pins with one eye, kept immovable. If the eye is about f or more from the nearer pin you will probably be unable to tell which pin is nearer to the eye, since a single eye is a very poor range-finder. Of course, if the eye is moved sideways, long experience enables you to tell at once, by using the methods of parallax, which is the nearer pin, by watching how the pins seem to move relatively to one another.

Since the single fixed eye can focus both pins at once, it can similarly focus simultaneously both images of the arms of the cross, and thereby leads us to infer (wrongly) that there is only one such image. What is perhaps the most remarkable fact is that the brain can entirely ignore all the images that are not in sharp focus, which Exp. 26 shows to lie all along the line between lens and eye. It is presumably a power developed by long practice, starting in the cradle. The parallax method, used here, enables us to locate the sharp images separately, undisturbed by the existence of blurred images at right angles to them; this power is not needed in everyday life, but it is useful in a laboratory.

Exp. 28. *Focal length of a compound lens, I*

For many practical purposes, such as the eyepieces and objectives of telescopes and microscopes and for cameras, it is the normal practice to use compound lenses, consisting of several lenses either in contact or mounted at some distance apart.

The theoretical approach to their behaviour, such as is given in the appended note and in Appendix B, appeals chiefly to mathematicians; even for them, and certainly for others, the first approach is best made entirely by experiment, and it is possible to go a considerable distance by experiment alone.

To simplify the first steps we will deal only with real images, and so avoid the complications of sign conventions; here and later the convention will be that $\frac{1}{f} = \frac{1}{u} + \frac{1}{v}$, f being positive for a convergent lens.

We will first investigate only the simplest form of compound lens, consisting of two thin convergent lenses (whose focal lengths will be denoted by f_1 and f_2), mounted at a distance (a cm.) apart, which is less than either f_1 or f_2. The first aim is to discover how the behaviour of this compound lens depends on f_1, f_2 and a.

Select two convergent lenses of nominal focal lengths between 10 and 20 cm., say $f_1 = 15$ cm. and $f_2 = 20$ cm. Measure by any method the values of f_1 and f_2, and set them up with their centres at A_1 and A_2, near the middle of the optical bench, at any convenient distance, say 10 cm., apart. Note the scale-readings of A_1 and A_2.

Set up a plane mirror to the right of A_2 and close to it, and by a pin and the parallax method locate F_1, the point where the pin coincides with its real image formed by two passages through the compound lens and one reflection at the mirror. In these experiments the alternative method can equally well be used, of taking as object a window cut in an opaque screen, with or without wire gauze across the window, lit by a lamp behind it, and receiving the image on a screen. More accurate results are obtainable by the former method, but with the latter it is easier to see what is going on, and for the moment this is the more important matter.

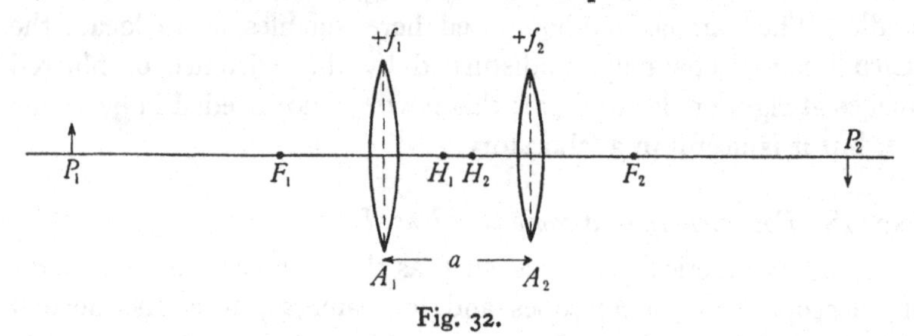

Fig. 32.

Then, as in the case of the single lens, F_1 must be the point at which a beam of light, coming from the right with all its rays parallel to the axis, will converge, so that we are justified in calling F_1 a Principal Focus of the compound lens. Note its scale-reading.

Similarly get the scale-reading of the other Principal Focus F_2 on the right of the lens.

Removing the mirror, set up a luminous object at some point P_1 to the left of F_1, and locate P_2 the image of P_1, produced on a screen by the compound lens; note the scale-readings of P_1 and P_2, and calculate the lengths of P_1F_1 and P_2F_2; it is advisable to adjust P_1 so that P_1F_1 and P_2F_2 do not differ much from one another. Calculate the value of $P_1F_1 \times P_2F_2$; denote it by X^2; from tables of square roots find X.

Repeat for two other positions of P_1. The resulting values of X will probably be nearly the same; if so, take their arithmetic mean.

When, in Exp. 21, we were dealing with a single thin lens, we found that $P_1F_1 \times P_2F_2$ equalled the square of the principal focal length of the single lens, in the ordinary meaning of that term.

We can if we like pursue that analogy and decide to call X 'the principal focal length of the compound lens', but it remains to be seen whether there is any sense in doing so. The term 'principal focal length' can mean either the length of the principal focus (since the principal focus is a point, this would be absurd) or the distance of the principal focus from some other point. There are two points distant X from F_1; take the point which is nearer to the lens than F_1, and call it H_1. Calculate the scale-reading of H_1. Similarly calculate the scale-reading of H_2, distance X from F_2; in the case of this compound lens it will be found that H_1 and H_2 do not coincide with one another or with A_1 or A_2. Nevertheless, it is possible that they may be suitable starting-points from which to measure the distances u and v of conjugate foci when we want to use the ordinary lens formula, just as F_1 and F_2 are suitable starting-points when we want to use Newton's lens formula.

This possibility can be tested from the observations already made; calculate from the scale-readings the lengths of P_1H_1 and P_2H_2 for each of the observed positions of P_1; calculate in each case the value of $\frac{1}{P_1H_1}+\frac{1}{P_2H_2}$; you will probably find that each of them is nearly equal to $1/X$.

If so, for this particular compound lens the ordinary lens formula $\frac{1}{u}+\frac{1}{v}=\frac{1}{X}$ holds good, provided that u and v represent the distances of object and image from H_1 and H_2 respectively, and X the focal length is obtained from $P_1F_1 \times P_2F_2 = X^2$.

Hence the use of the term focal length is fully justified; the points H_1 and H_2, which have fixed positions for any particular compound lens, are so important that they are termed the Principal Points of the lens.

Further, the letter X can now be replaced by the customary letter f, and the ordinary lens formula for a compound lens can be written as $\frac{1}{u}+\frac{1}{v}=\frac{1}{f}$, and Newton's lens formula as $P_1F_1 \times P_2F_2 = f^2$.

Note on the Theory

Consider two thin lenses, at A_1 and A_2, a apart, with focal lengths of f_1 and f_2 (which may have either sign). Suppose that the A_1 lens produces an image, real or virtual, at Q of an object at P_1, and that the A_2 lens produces an image at P_2 of Q.

Appendix B (i) and (ii) gives a complete theoretical analysis of the relations between the positions of P_1, Q and P_2 when the lenses are of any thickness; we can adapt that analysis to this particular case, including Fig. 33 and the use of the 'real-is-positive' sign convention, by putting $t=t'=0$ and changing f into f_1, f' into f_2, F into f, P_0 into P_1, P_2 into P_1, L_1 into H_1, L_2 into H_2, $-U$ into u_1 and V into u_2.

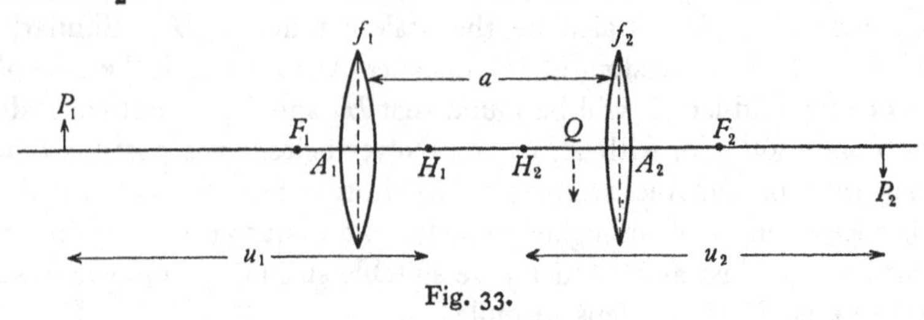

Fig. 33.

Then from (6) in Appendix, A_1, A_2, H_1 and H_2 coalesce into A_1 in Fig. 33; from (7) A_1', A_2', H_1' and H_2' coalesce into A_2 in Fig. 33; from (10) C becomes a; from (12)

$$\frac{1}{f}=\frac{1}{f_1}+\frac{1}{f_2}-\frac{a}{f_1 f_2}. \tag{1}$$

and, if we wish to use Halley's lens-formula, from (13)

$$A_1H_1=d=\frac{af}{f_2}=\frac{af_1}{f_1+f_2-a}, \tag{2}$$

and

$$A_2H_2=-d'=\frac{af}{f_1}=\frac{af_2}{f_1+f_2-a}, \tag{3}$$

and from (14)

$$\frac{1}{u_1}+\frac{1}{u_2}=\frac{1}{f_1}+\frac{1}{f_2}-\frac{a}{f_1 f_2}, \tag{4}$$

where u_1 and u_2 are measured from the principal points of the compound lens, H_1 and H_2 respectively, signs being assigned according to the 'real-is-positive' sign convention so that f_1 and f_2 are positive if the respective lenses are convergent, and vice versa.

If u_1 becomes infinitely large, u_2 becomes the distance from H_2 of the Principal Focus F_2 of the combination, as ordinarily defined; if then we denote by f the Principal Focal Length of the combination, being the distance between H_2 and F_2,

$$\frac{1}{f}=\frac{1}{f_1}+\frac{1}{f_2}-\frac{a}{f_1 f_2}. \tag{5}$$

Similarly, we see that there is another principal focus F_1, distant f from H_1, where

$$\frac{1}{f}=\frac{1}{f_1}+\frac{1}{f_2}-\frac{a}{f_1 f_2}.$$

Hence, from (4), $$\frac{1}{u_1}+\frac{1}{u_2}=\frac{1}{f}, \qquad (6)$$

where $$\frac{1}{f}=\frac{1}{f_1}+\frac{1}{f_2}-\frac{a}{f_1 f_2}$$
$$=\frac{f_2+f_1-a}{f_1 f_2},$$

so that $$f=\frac{f_1 f_2}{f_1+f_2-a}. \qquad (7)$$

It should be noted that the two principal foci F_1 and F_2 are distant f, or by (7) $\frac{f_1 f_2}{f_1+f_2-a}$, from H_1 and H_2 respectively; since from (2)

$$A_1H_1=\frac{af_1}{f_1+f_2-a} \quad \text{and} \quad F_1H_1=\frac{f_1 f_2}{f_1+f_2-a}, \quad F_1A_1=\frac{f_1(f_2-a)}{f_1+f_2-a}.$$

Similarly, $$F_2A_2=\frac{f_2(f_1-a)}{f_1+f_2-a}.$$

Hence if a is less than f_2, F_1 will lie outside the combination, and we can locate it by a search pin or luminous object and screen, as is required to be done in the experiment; similarly with F_2; so that the range of a for which this experiment is workable is from zero to the smaller of the two focal lengths f_1 and f_2, as is mentioned there.

If a is increased beyond f_2, F_1 will disappear into the combination and this particular experiment cannot be completely carried out; from (7) we see that if f_1 and f_2 are unchanged and both are positive, f increases as a decreases, so H_1 moves to the right more rapidly than F_1. In particular, when $a=f_1+f_2$, f becomes infinite and H_1 goes to an infinite distance; this case is dealt with in Exp. 32. But during these changes P_1 and P_2 may remain outside the combination, so that the combination may continue to act as a convergent lens, giving a real image of a real object, even when a is larger than f_1+f_2, and f becomes negative, in accordance with (7). It is easy to verify this with the set-up of this experiment.

Exp. 29. *Focal length of a compound lens, II*

Exp. 28 showed the existence and properties of the principal foci, principal points and focal length of a compound lens, and the use of the ordinary and Newton's lens formulas in one particular case. The next step is to find out experimentally how the focal length f of a compound lens depends in general on the focal lengths, f_1 and f_2, of the constituent lenses and the distance a between them.

The way in which f depends on a can be discovered experimentally by measuring f for a particular pair of lenses when the distance a is varied. The observations made in Exp. 28 with a particular pair of lenses furnish one pair of values of X, or f, and a; use the same lenses at a different distance a_1 cm. apart and find the corresponding value of X, or f. For this, we must locate F_1 and F_2, and at least one pair of conjugate foci, P_1 and P_2, and calculate f from the relation $P_1F_1 \times P_2F_2 = f^2$, as in that experiment; it is better to get two pairs of conjugate foci, and take the arithmetic mean of the two resultant values of f.

Repeat for two other values of a, and plot a graph connecting the values of $1/f$ and a.

It is probable that your graph will be a straight line; if so, there is a linear numerical relation between $1/f$ and a which can safely be assumed to hold good for these two lenses for all values of a within a certain range, and which may hold good for all values of a. Let us express this in the form

$$\frac{1}{f} = A + Ba, \tag{1}$$

where A and B do not involve a.

Putting $a=0$ it will be seen from (1) that the intercept on the $1/f$ axis is A; putting $1/f=0$ it will be seen that the intercept on the a axis is $-A/B$. So you can calculate the numerical values of A and B for this pair of lenses from the observed intercepts of your graph. It can safely be assumed that (1) will apply to any other pair of thin lenses, though the values of A and B will be different.

The next step is to find how A and B depend on f_1 and f_2. This could be done, quite satisfactorily, by keeping one lens (f_1 say) and a the same, and varying f_2 by using a number of different lenses. We could thus get a number of pairs of values of f and f_1, and if we plotted a graph connecting $1/f$ and $1/f_2$ it would probably turn out to be linear. From its intercepts we could find a relation between f and the focal length of a constituent lens, and combining this with (1) we could find the complete formula for $1/f$.

But this involves a good deal of experiment, and requires a number of lenses of measured focal lengths. It is much quicker, and almost as satisfactory, to use the following indirect method of reaching the same result from the graph which you have already obtained.

A and B in (1) do not depend on a, so we can take any value we please for a, on the assumption that (1) holds good for all values of a.

Imagine then that $a=0$; (1) then becomes $1/f=A$ for a combination consisting of two thin lenses in contact. You probably know from earlier laboratory work with thin lenses that *under these conditions*

$$\frac{1}{f}=\frac{1}{f_1}+\frac{1}{f_2}, \quad \text{so that} \quad A=\frac{1}{f_1}+\frac{1}{f_2}. \tag{2}$$

Comparison between the intercept of your graph on the a axis and the known values of f_1 and f_2 makes it obvious that, in this case at least, the intercept equals f_1+f_2. Now this intercept has been shown to equal $-A/B$; hence, substituting the known value for A deduced from (2), the intercept equals $-\frac{1}{B}\left(\frac{1}{f_1}+\frac{1}{f_2}\right)$. If it also equals f_1+f_2 we have

$$f_1+f_2=-\frac{1}{B}\left(\frac{1}{f_1}+\frac{1}{f_2}\right),$$

or

$$B=-\frac{1}{f_1+f_2}\left(\frac{1}{f_1}+\frac{1}{f_2}\right)=-\frac{1}{f_1f_2},$$

and the general relation which we set out to find is

$$\frac{1}{f}=\frac{1}{f_1}+\frac{1}{f_2}-\frac{a}{f_1f_2}, \tag{3}$$

or

$$f=\frac{f_1f_2}{f_1+f_2-a}. \tag{4}$$

The theoretical justification for this is given at (7) in the note appended to Exp. 28.

Exp. 30. *Principal points of a compound lens*

In order to use for practical purposes the ordinary lens formula $\frac{1}{u}+\frac{1}{v}=\frac{1}{f}$ with a compound lens in which we know the focal lengths, f_1 and f_2, of the two lenses and a, the distance between them, we must have not only the general formula for f in terms of a, f_1 and f_2, which we found in Exp. 29 to be $f=\frac{f_1f_2}{f_1+f_2-a}$, but also general formulas for the positions, relative to the lens, of the two principal points H_1 and H_2 from which the distances u and v of conjugate foci are measured.

These can be found as follows; all the necessary observations were made in Exps. 28 and 29, and it will be assumed that they are available; if not, they should be made as described there.

Using a pair of lenses with known values of f_1 and f_2, for one value of a, you found in Exp. 28 the scale-reading of H_1, and noted the scale-reading of A_1; you can hence deduce the length A_1H_1. Using the observations made in Exp. 29 with other values of a, deduce in the same way the corresponding values of A_1H_1. Plot a graph connecting $1/A_1H_1$ and $1/a$.

If as is probable this is a straight line, find its intercepts on the two axes. What we have to discover is the connection between the numerical values of these intercepts and the numerical values of f_1 and f_2. It is not likely to be obvious, but we have to grope about until we find it. Since we are using the reciprocals of a and A_1H_1 it is a good plan to find the reciprocals of the intercepts; when this is done it becomes fairly obvious that the reciprocal of the intercept on the $1/A_1H_1$ axis equals $-f_1$, and the reciprocal of the intercept on the $1/a$ axis equals f_1+f_2.

If this is true in general, the relation between A_1H_1 and a must be

$$\frac{1}{A_1H_1}=\frac{f_1+f_2}{f_1}\frac{1}{a}-\frac{1}{f_1}, \tag{1}$$

since this equation gives these intercepts.

Equation (1) can be rearranged as

$$\frac{1}{A_1H_1}=\frac{f_1+f_2-a}{af_1} \quad \text{or} \quad A_1H_1=\frac{af_1}{f_1+f_2-a}. \tag{2}$$

Proceeding similarly for the other principal point H_2 we get

$$A_2H_2=\frac{af_2}{f_1+f_2-a}. \tag{3}$$

These are the general formulas we set out to find. We have actually arrived at them by the use of only one pair of lenses, but their truth for any other values of f_1 and f_2 can be checked, if thought advisable, by applying them to any other pair of lenses.

Since $F_1A_1=f-AH_1=f-\dfrac{af_1}{f_1+f_2-a}$ by (2), and $f=\dfrac{f_1f_2}{f_1+f_2-a}$ by the results of Exp. 29, we get $F_1A_1=\dfrac{f_1(f_2-a)}{f_1+f_2-a}$. Similarly,

$$F_2A_2=\frac{f_2(f_1-a)}{f_1+f_2-a}.$$

These formulas show that if a exceeds f_1 or f_2 the corresponding principal focus lies inside the compound lens, so that it cannot be located by the plane mirror method, which is an essential step in Exps. 28 and 29. The formulas are obtained theoretically in the note appended to Exp. 28.

Exp. 31. *Camera lens giving variable magnification*

The size of the image of a distant object on the film of a camera with a lens of fixed focal length is directly proportional to that focal length; if a larger image is required, either the lens must be changed, or a concave lens (often called an 'enlarger') can be put in front of the convex lens. In telephoto cameras this latter is usually a permanent fitting, mounted so that the distance between the two lenses can be adjusted as required. We can investigate the relation between this distance and the size of the image by the following experiments.

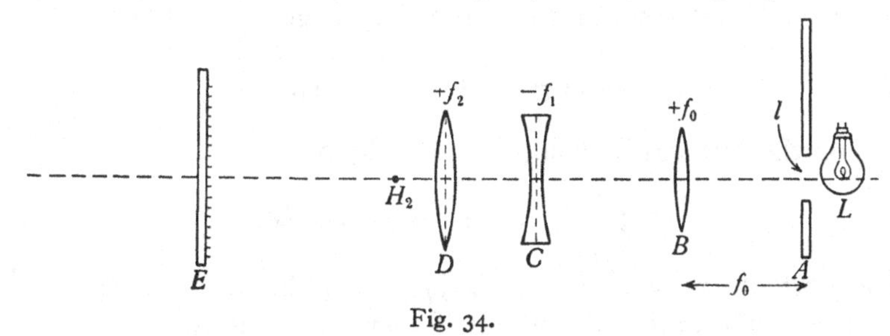

Fig. 34.

The most convenient way of producing the effect of a very distant luminous object on a laboratory bench is to set up a piece of cardboard, as at A in Fig. 34, with a rectangular window cut in it, with vertical sides l cm. apart, and to put behind it, at L, a lamp; this should be 'pearl' or 'opal', or if it is of clear glass a piece of translucent paper should be fixed over the window. A convergent lens, focal length f_0 cm., should be set up at B, so that $BA = f_0$; this is ensured by putting a plane mirror beyond B and adjusting until a sharp image of the window is visible on the cardboard beside the window.

Next set up at C a concave lens, of focal length $-f_1$ cm.; the length CB is theoretically unimportant, but better illumination of the final image is secured if CB is small, such as 10 cm. C is to remain unchanged throughout the experiment. At D set up a convex lens, of focal length $+f_2$ cm., and at E a screen with a horizontal millimetre scale fixed on it (preferably printed on white paper); instead of a scale, a pair of dividers may be used for measuring the breadth of the images on the screen.

f_1 should be greater than f_2, say about 15 and 10 cm. respectively

Denote by x cm. the distance DC between the two camera lenses. Start with x as small as the lens-holders will permit, measure x_1 and move the screen, which represents the film of the camera, until the window is in the sharpest possible focus on it; measure accurately on the scale, or by the dividers, the width, m cm., of the image of the window.

Repeat for a succession of values of x, and plot a graph connecting $1/m$ with x.

If, as is probable, this is a straight line, find its intercept on each axis.

The appended note shows that theoretically the relation between $1/m$ and x is

$$\frac{1}{m}=\frac{f_0}{l}\left\{\frac{x}{f_1 f_2}-\frac{1}{f_1}+\frac{1}{f_2}\right\}. \tag{1}$$

Hence the intercept on the x axis should theoretically be

$$f_1 f_2\left(\frac{1}{f_1}-\frac{1}{f_2}\right) \quad \text{or} \quad f_2-f_1,$$

and on the $1/m$ axis it should theoretically be

$$\frac{f_0}{l}\left(\frac{1}{f_2}-\frac{1}{f_1}\right) \quad \text{or} \quad \frac{f_0}{l f_1 f_2}(f_1-f_2).$$

Substitute the known values of f_0, l, f_1 and f_2 (they must be measured by any of the ordinary methods if they are not already known) in these two expressions, and compare them with the intercepts of your graph, thereby checking (1).

In carrying out the experiment you will probably have noticed that, as the distance x between the lenses is steadily increased the size of the image steadily decreases. This is in agreement with (1), since the expression within the bracket increases with x, and therefore m decreases as x increases. It will be seen that this expression within the brackets is the same (when the sign of f_1 has been duly changed, since this lens is now concave) as the expression on the right-hand side of (3) in Exp. 29, which was there shown experimentally to be equal to $1/f$, where f is the equivalent focal length of the combination. Hence (1) can be written

$$\frac{1}{m}=\frac{f_0}{l}\frac{1}{f} \quad \text{or} \quad m=\frac{l}{f_0}f,$$

and we reach the usual result that the size of the image on the film is directly proportional to the equivalent focal length of the camera lens, whether it be simple or compound.

Presumably then we could equally well have made up our variable focus combination with two convex lenses instead of one convex and one concave lens. It is true that there would be one difference of behaviour between these two combinations; since in the former case $\frac{1}{f}$ is $-\frac{x}{f_1 f_2}+\frac{1}{f_1}+\frac{1}{f_2}$, f will *increase*, and so will the size of the image, as x increases, whereas in the latter case $\frac{1}{f}$ is $+\frac{x}{f_1 f_2}-\frac{1}{f_1}+\frac{1}{f_2}$ and therefore f *decreases*, as will the size of the image, as x increases. But the important practical feature is that in both cases the equivalent focal length, and hence the size of the image, changes continuously with x.

There is, however, a serious practical drawback to the use of two convex lenses, which you can easily discover for yourself by substituting a convex lens of, say, +15 cm. focal length for the −15 cm. one used in the foregoing. You will probably find, as is shown theoretically above, that as x is increased the size of the image is increased, but that in spite of the increase of f the lens D gets nearer to the film E, and actually meets it before the image is much increased in size. This is a consequence of the fact that when the film is in focus it is at a distance f from the principal point H_2 of the combination, and not from D; if both lenses are convex, H_2 lies between D and C; and D moves to the left so much faster than H_2 that its movement outweighs the increase in f, and so it closes up on E. When, on the other hand, there is a concave lens at C, the principal point H_2 is nearer to E than D, as in Fig. 34, so that however small f becomes D cannot close completely on to E.

Hence it is best to use the combination of one convex and one concave lens.

Note on the Theory

In Fig. 35, a beam diverging from P_0, l cm. above the axis at A, emerges from the lens B as a parallel beam with its rays making an angle β with the axis, where

$$\beta=\frac{l}{f_0}. \tag{1}$$

After passing through lens C this parallel beam diverges from a point P_1, m_1 cm. say above F_1, on the principal focal plane of C, and therefore f_1 cm. from C and f_1+x from D. Hence

$$\frac{m_1}{f_1}=\beta, \tag{2}$$

where m_1 is the size of the virtual image formed by C.

Hence if the beam is caused by D to converge to a point P_2 on E, m cm. say below the axis,

$$\frac{1}{DF_1}+\frac{1}{DE}=\frac{1}{f_2} \quad \text{or} \quad \frac{1}{f_1+x}+\frac{1}{DE}=\frac{1}{f_2}. \tag{3}$$

And as the ray through the centre of lens D is undeflected

$$\frac{m}{DE}=\frac{m_1}{f_1+x}. \tag{4}$$

But from (1) and (2) $\quad m_1=f_1\beta=\frac{lf_1}{f_0},$

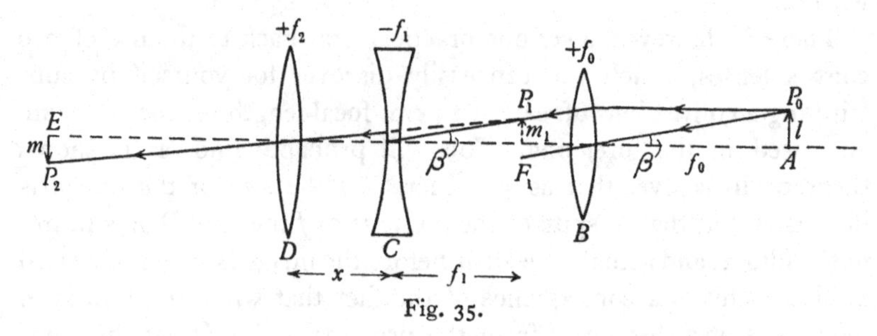

Fig. 35.

and from (3) and (4)

$$\frac{1}{f_1+x}+\frac{m_1}{m}\frac{1}{f_1+x}=\frac{1}{f_2}.$$

Hence

$$\frac{1}{f_1+x}+\frac{lf_1}{f_0 m}\frac{1}{f_1+x}=\frac{1}{f_2},$$

or

$$f_1+x=f_2\left(1+\frac{lf_1}{f_0 m}\right),$$

or

$$x=\frac{lf_1 f_2}{f_0}\frac{1}{m}+f_2-f_1.$$

In the experiment this is used in the form

$$\frac{1}{m}=\frac{f_0}{l}\left\{\frac{x}{f_1 f_2}-\frac{1}{f_1}+\frac{1}{f_2}\right\}.$$

Exp. 32. *Telescopes*

It will be seen that the formulas found in Exps. 28 and 29 have no practical value if the distance a between the lenses is equal to the sum of f_1 and f_2, since the focal length of the compound lens and the distances of its principal foci and principal points from the lens then become infinitely large. In the astronomical telescope as a whole, and in the erecting lenses of a terrestrial telescope, this condition holds good, so it is worth while to investigate the behaviour of this special form of compound lens. It can be done to some extent by direct experiment, as follows.

Take two convergent lenses, of focal lengths f_1 and f_2; it is convenient to have f_1 and f_2 fairly small, about 15 and 10 cm. respectively. Measure f_1 and f_2 by any means; set up the weaker (f_1) with its centre at A_1 at about the middle of the optical bench, and the stronger (f_2) with its centre at A_2, f_1+f_2 from A_1; note the scale-readings of A_1 and A_2.

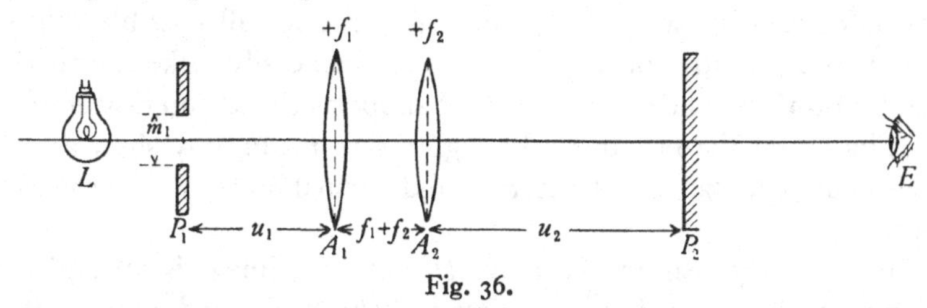

Fig. 36.

For the experiment the most suitable object, to be set up at P_1, is a fairly narrow rectangular window in a screen, with vertical sides about 2·5 cm. long and all its edges sharply cut, covered with thin translucent paper and lit by a lamp behind it, the image being focused on a screen at P_2; measure accurately the height of the vertical window, and denote it by m_1. Denote P_1A_1 by u_1, and P_2A_2 by u_2; note that u_1 and u_2 are not measured from the principal points of the compound lens, since these points are infinitely far away. Keeping A_1 and A_2 undisturbed, get six or seven pairs of conjugate focal lengths u_1 and u_2, extending over the maximum attainable range; at each setting of P_2 measure accurately, with the help of a pair of dividers, the vertical height (m_2) of the image of the window.

Plot a graph connecting u_1 and u_2. If, as is probable, this is a straight line, find its intercepts on the two axes.

The appended note shows that the intercepts on the u_1 and u_2 axes should theoretically be $f_1^2\left(\frac{1}{f_1}+\frac{1}{f_2}\right)$ and $f_2^2\left(\frac{1}{f_1}+\frac{1}{f_2}\right)$ respectively. Compare the intercepts of your graph with these values; if they agree, the relation between u_1 and u_2 must be

$$\frac{u_1}{f_1^2}+\frac{u_2}{f_2^2}=\frac{1}{f_1}+\frac{1}{f_2}, \qquad (1)$$

since this represents a straight line with these intercepts.

The appended note shows also that the size of the image should theoretically be the same for all values of u_1, the ratio m_2/m_1 being equal to the ratio f_2/f_1. Compare your experimental result with this theoretical one.

The graph and the equation (1) derived from it show that, if u_1 exceeds $f_1^2\left(\frac{1}{f_1}+\frac{1}{f_2}\right)$, u_2 becomes negative, so that the image at P_2 is no longer real and receivable on the screen. Hence experiments on the foregoing lines cannot give us the value of u_2 under such conditions and thereby check the complete truth of (1), which the appended note shows to be theoretically true for all possible values of u_1 and u_2, positive or negative. But you can easily make an experiment on different lines which at least supports this theoretical result. It is based on the fact that, although a screen can only show a real image, an eye can see both real and virtual images, if properly placed.

Put P_1 fairly near to A_1, move P_2 until the image is focused on it, put your eye at E some distance behind P_2, and remove the screen P_2. You will then see the real image of P_1. Move P_1 steadily away from A_1 and watch the image; there will be no apparent break in continuity as the image passes through A_2 and changes from real to virtual. You cannot, on a laboratory bench, continue moving P_1 to an infinite distance, but you can produce the same effect by setting up a convex lens in front of P_1, so that P_1 is at its principal focus. You will then find that, even when P_1 is at an infinite distance from A_1, you can still see the virtual image, and can now bring your eye close to A_2 without the image becoming blurred, because P_2 has also moved off to an infinite distance from A_2. This compound lens then functions as an astronomical telescope.

Since there is no break in continuity when the image becomes virtual it is at least probable that the law governing the size of the image continues to hold when the image is virtual. If so, the image should remain of the same size, f_2/f_1 times the size of the object and therefore in this case smaller than the object. The note shows that this supposition is theoretically sound.

Now in an astronomical telescope f_2, the focal length of the eye-piece, is always many times smaller than f_1, the focal length of the objective, so the image formed by an astronomical telescope should always be much smaller than the object. And yet practical experience with an astronomical telescope shows that it acts as a magnifier, not a reducer (provided that you do not look through the wrong end).

The explanation of this apparent contradiction may not be obvious at once; it is certainly not that the observations or the conclusions drawn from them were incorrect. The contradition arises from the common habit of treating all points which are 'at an infinite distance' from the observer as though they were at the same distance from him.

In this particular case, equation (1) shows that

$$-\frac{u_2}{u_1}=\frac{f_2^2}{f_1^2}-\frac{f_2^2}{u_1}\left(\frac{1}{f_1}+\frac{1}{f_2}\right);$$

when u_1 is very large, the second term on the right-hand side vanishes, in comparison with the first, and $-\frac{u_2}{u_1}=\frac{f_2^2}{f_1^2}$. So the distance

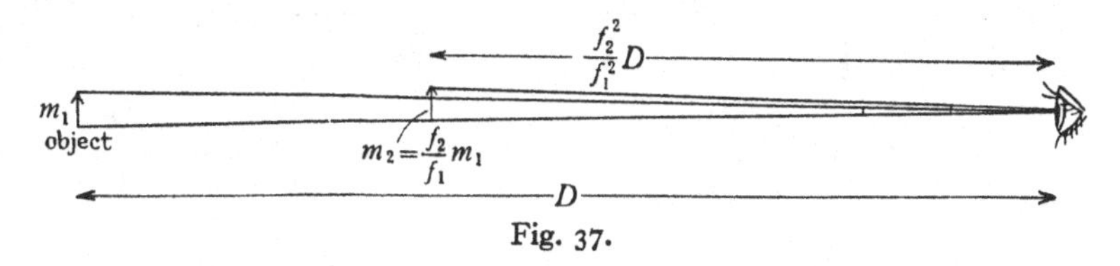

Fig. 37.

of the image from the eye is then f_2^2/f_1^2 times the distance of the object from the eye. This relative nearness (f_2 is smaller than f_1) more than counterbalances the fact that, as we have seen, the image is f_2/f_1 times the size of the object; Fig. 37 shows that if the object is D from the eye, it subtends an angle m_1/D at the eye, whereas the image subtends at the eye an angle $\frac{f_2}{f_1}m_1\Big/\frac{f_2^2}{f_1^2}D$ or $\frac{f_1}{f_2}\frac{m_1}{D}$ or $\frac{f_1}{f_2}$ times the angle subtended by the object. Now f_1/f_2 is greater than 1·0, so the image *looks* larger than the object, although it is in fact smaller.

It is possible, though not altogether easy, to arrange and carry out an experiment to check roughly these calculations, and it is worth while to do so because even a rather clumsy experiment may be more illuminating than very precise and neat calculations.

The screen P_2 and the lamp behind the window in Fig. 36 will not be needed and the experiment should be performed in a fully lit room. Put the object P_1 at one end of the scale and a convex lens at A_3, at such a distance from P_1 that P_1 is at the principal focus of this lens; this ensures that the object is effectively at a very great distance.

Take for the eye-lens (A_2 in Fig. 38 which is a vertical section) a short focus lens, say $f_2 = 10$ cm., and for the objective A_1 a long focus lens, say $f_1 = 30$ cm.; put A_2 at the other end of the bench from P_1, and A_1 at $f_1 + f_2$ from A_2. The focal length of A_3 should be small enough to enable you to get your head between A_1 and A_3.

Fix in a retort stand a pair of dividers, so that the points are in a horizontal line a good deal higher than the optic axis and at right angles to the plane of the paper in Fig. 38, above Q_1 which is, say, 50 cm. from A_3. If you put one eye close to the lens A_3, it will see the image of the window P_1; with the other eye, about vertically above the former, look at the points of the dividers, and open or

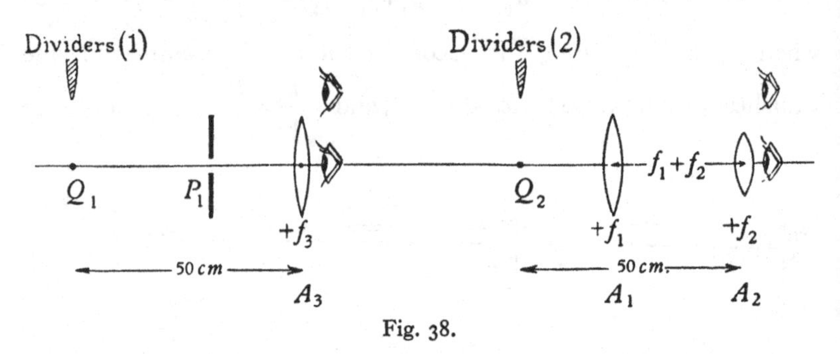

Fig. 38.

close them until they can be brought to coincide simultaneously with the edges of the window as seen by the former eye. This operation gives scope for individual skill, and cannot be done with great accuracy. Measure the distance between the points of the dividers.

Next, move the dividers in their retort stand to a point Q_2, the same distance from A_2 as Q_1 was from A_3, and again well above the optic axis. Repeat the operation, and calculate the ratio between the two distances between the dividers' points. This must equal the ratio of the magnitude of the distant object, as seen by the naked eye, to the magnitude of the image as seen through the telescope; more strictly, it is the ratio of the angle subtended by the object at the eye to the angle subtended by the image; and it should, as we have seen, be equal to f_2/f_1. A fair agreement between practice and theory is all that can be expected from such an experiment.

A numerical example may help to collect and clarify the foregoing points. Suppose you have a telescope with an objective of focal length $f_1 = 5$ ft., and an eyepiece of focal length $f_2 = 3$ in. (so that

$f_2/f_1 = \frac{1}{20}$), and direct it to the sun, whose distance is about 92,000,000 miles. The image formed by the telescope will be $\frac{f_2^2}{f_1^2} \times 92{,}000{,}000$ or 230,000 miles away; the moon is about 240,000 miles away, so the image of the sun will be about as far away as the moon. The sun and moon both subtend nearly the same angle at the naked eye; the image of the sun subtends f_1/f_2 times this angle, so the image of the sun in this telescope will *seem* to be 20 times the size of the moon as seen without a telescope.

Note on the Theory

Using Fig. 36, suppose that the lens at A_1 produces an image, v cm. from A_1, of the object at P_1. Then $\frac{1}{u_1} + \frac{1}{v} = \frac{1}{f_1}$, and hence

$$v = \frac{u_1 f_1}{u_1 - f_1}. \qquad (1)$$

This intermediate image is $f_1 + f_2 - v$ from the lens A_2, so that

$$\frac{1}{f_1 + f_2 - v} + \frac{1}{u_2} = \frac{1}{f_2},$$

and hence

$$f_1 + f_2 - v = \frac{u_2 f_2}{u_2 - f_2}. \qquad (2)$$

Adding (1) and (2),

$$f_1 + f_2 = \frac{u_1 f_1}{u_1 - f_1} + \frac{u_2 f_2}{u_2 - f_2},$$

or

$$f_1 - \frac{u_1 f_1}{u_1 - f_1} = \frac{u_2 f_2}{u_2 - f_2} - f_2,$$

or

$$\frac{-f_1^2}{u_1 - f_1} = \frac{f_2^2}{u_2 - f_2}.$$

Inverting,

$$-\frac{u_1 - f_1}{f_1^2} = \frac{u_2 - f_2}{f_2^2} \quad \text{or} \quad \frac{\mathbf{u_1}}{\mathbf{f_1^2}} + \frac{\mathbf{u_2}}{\mathbf{f_2^2}} = \frac{f_1}{f_1^2} + \frac{f_2}{f_2^2} = \frac{\mathbf{1}}{\mathbf{f_1}} + \frac{\mathbf{1}}{\mathbf{f_2}}. \qquad (3)$$

The intercepts of the graph represented by (3) will be $f_1^2\left(\frac{1}{f_1} + \frac{1}{f_2}\right)$ on the axis of u_1, and $f_2^2\left(\frac{1}{f_1} + \frac{1}{f_2}\right)$ on the axis of u_2.

No assumptions have been made here as to the signs of u_1, u_2, f_1 or f_2.

The two lenses are f_1+f_2 apart, so they have a common principal focus, at F say. A ray P_1R_1 from the top of the object parallel to the axis passes through F and emerges along R_2P_2, parallel to the axis. Hence, whether the image is real or virtual its top P_2 lies on R_2P_2 (see Fig. 39).

By similar triangles,

$$\frac{f_1}{f_2}=\frac{R_1A_1}{R_2A_2}=\frac{m_1}{m_2}, \quad \text{or} \quad \mathbf{m_2}=\frac{\mathbf{f_2}}{\mathbf{f_1}}\mathbf{m_1}.$$

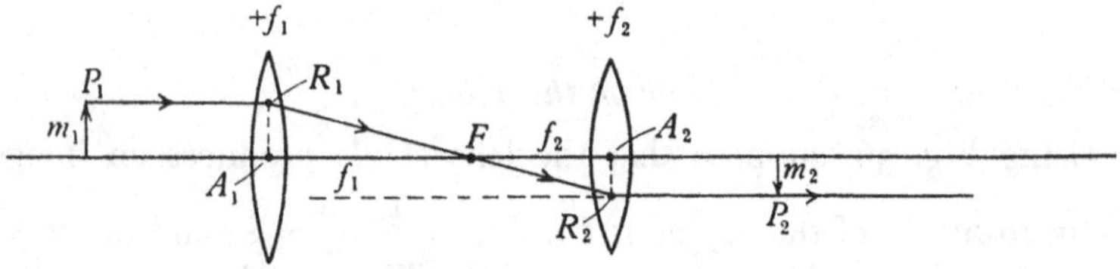

Fig. 39.

Exp. 33. *Thick lenses in contact*

In the foregoing experiments on lenses many cases have arisen where the thickness of the lens affects the result and must be allowed for. The mathematical theory of thick lenses when forming combinations such as are used in nearly all practical optical instruments (see Appendix B) is long and complicated, though some of its results are simple. The following experiment uses simple means to illustrate, and to verify roughly, one such result; it uses an almost extravagantly thick lens, so that the effect of thickness may be large enough to be easily measurable.

Set up a compound lens consisting of four or five lenses, all of say 5 cm. diameter, with the centres of their faces in contact, bedded down on a lump of plasticine, with another lump fixed on top of them to keep them steady. The lenses must be chosen so that the compound lens is convergent, i.e. the sum of the reciprocals of their focal lengths must be positive; for example, they may have focal lengths of about +10, −15, +20, +15 and −15 cm.; this combination would have a thickness t along the axis of about 2·3 cm. which you should measure with callipers, and a focal length of about 12 cm.

Set up the compound lens on the optical bench. By means of a pin and a good plane mirror locate by the direct parallax method its two principal foci F_1 and F_2, and note their scale-readings. It is essential that the positions of F_1 and F_2 should be found with the greatest possible precision; the mean of not less than five or six independent settings should be taken in each case, and the search pin should be set so that it and the image appear to cling together for about one-fifth of the traverse across the lens, but at each end of the traverse the pin should appear to move faster than the image (see Appendix A). The success of this experiment depends largely on the correctness of these two determinations.

The next step is to determine the positions of one pair of conjugate foci, C_1 and C_2. Remove the pins from F_1 and F_2, but keep the lens unmoved; set up an object pin at C_1, about half of F_1F_2 on the left of F_1; locate C_2, the conjugate focus of C_1, with the same care as was given to the location of F_1 and F_2; note the scale-readings of C_1 and C_2. Hence calculate the lengths C_1F_1 and C_2F_2 respectively, and the square root of their product; denote this by F, as that $C_1F_1 \times C_2F_2 = F^2$.

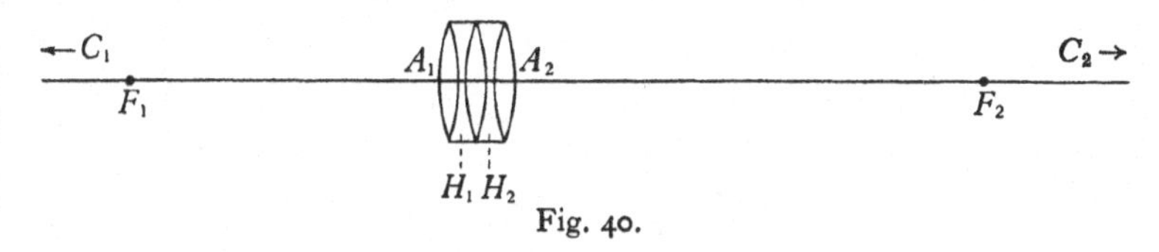

Fig. 40.

Repeat for two or three positions of C_1, several cm. apart, you will probably find that F is nearly the same in each case, so that Newton's lens-formula holds good here as is shown theoretically in Appendix B (ii); F will be the focal length of the combination; determine its value as the arithmetic mean of your results.

Hence, calculate (as in Exp. 28) the position of the points H_1, H_2 at distances of F from F_1 and F_2 respectively towards the lens; these are, as shown in Appendix B, the principal points of the lens which must be used as the bases from which the distances of the conjugate foci C_1 and C_2 must be measured when we use the ordinary lens-formula $1/v + 1/V = 1/F$; calculate the values of C_1H_1 and C_2H_2, or U and V, and determine the sum of their reciprocals in each case, you will probably find that (*a*) this sum is nearly constant, and (*b*) nearly equal to $1/F$ as already determined. If so, you will have verified (14) of Appendix B, from which it appears that Halley's lens-formula holds good for the compound lens.

Calculate the length H_1H_2 and compare it with t; if you have worked accurately you will probably find that H_1H_2 is roughly equal to $t/3$.

Finally, find the scale-readings of the centres, A_1 and A_2 in Fig. 40, of the faces of the lens; hence calculate the values of A_1H_1 and A_2H_2. These will probably turn out not to be equal to one another, as they are in a bi-convex lens (especially if one outer face of the compound lens is convex and the other concave), but it is certainly not worth while to attempt to compare them with their theoretical values, since that would involve several hours of work.

PART III. HEAT AND ELECTRICITY

Exp. 34. *Change of viscosity with temperature*

Exp. 14 showed how the viscosity of cold water can be measured; this experiment shows how we can investigate, with very simple apparatus, the changes in that viscosity as the temperature changes.

Take a glass tube, about 9 mm. external diameter and 25 cm. long, heat it at a point about 8 cm. from one end and draw it out into a *very* fine capillary tube. Cut the capillary at such a point that there is about 4 cm. of it attached to the longer tube. Clean this by sucking first alcohol and then water into it.

Lash three or four turns of cotton tightly round the uncontracted part of the tube at points *A*, *B* and *C* in Fig. 41, about 1, 4 and $5\frac{1}{2}$ cm. respectively from the point where it begins to narrow. Or use any other method of marking these points, such as rings cut off a rubber tube of a size to slip tightly over the glass tube.

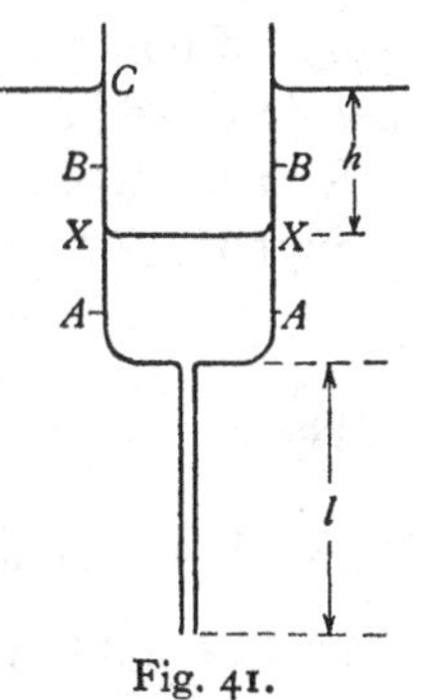

Fig. 41.

Closing the top with your finger (or, if preferred, by a rubber tube with a spring clip) clamp the tube vertically in a beaker of cold water so that *C* is in the free surface of the water. Remove your finger and measure the time taken by the water to rise from *A* to *B*. If this time is inconveniently long, shorten the capillary; if it is too short, move *B* and *C* up the tube. Read the temperature of the water.

Repeat the experiment with water of three different temperatures up to about 50° C., stirring the water throughout each experiment. Plot a graph of times as ordinates against temperatures as abscissae.

The appended note on theory shows that under these conditions the coefficient of viscosity of water at any given temperature is directly proportional to the time taken by water at that temperature to rise from *A* to *B*. Hence any ordinate of your graph, when multiplied by a suitable constant, should theoretically be the viscosity of water at the corresponding temperature; you can determine the value of this constant from the fact that the recognised value of the viscosity of water at 25° C. is 0·00893 g. cm.$^{-1}$ sec.$^{-1}$, by comparing the ordinate of your graph at 25° C. with that number. Hence you can easily deduce from your graph one giving directly the recognised values of the viscosity for all temperatures within its range.

Note on the Theory

First let us assume that the changes in the dimensions of the glass apparatus and in the density of water due to the rises in temperature are so small in comparison with the changes in viscosity due to the same rise in temperature that we may ignore them.

The coefficient of viscosity (η) of a liquid is measured (see Exp. 14) by $\frac{\pi p r^4 t}{8lV}$ g. cm.$^{-1}$ sec.$^{-1}$, where V c.c. is the volume of the liquid which flows in t sec. through a capillary tube of length l cm. and radius of bore r cm. when a pressure-difference p dynes per sq.cm. is maintained between the ends of the tube in excess of the pressure-difference needed to keep the liquid in the tube in equilibrium.

Suppose that the subscripts 0 and 1 are attached to these quantities when the temperatures are θ_0° C. and θ_1° C.

Then $$\eta_0 = \frac{\pi p_0 r_0^4 t_0}{8 l_0 V_0} \quad \text{and} \quad \eta_1 = \frac{\pi p_1 r_1^4 t_1}{8 l_1 V_1}.$$

But we have assumed that we can disregard changes in pressure-head, in r_1, in l and in V due to the rise in temperature; hence $\eta_1 = \eta_0 \frac{t_1}{t_0}$, since these are the only quantities affected by the change of temperature.

We should next inquire into the validity of this assumption, that is whether in this experiment we can ignore the effect of temperature on the dimensions of the apparatus and the density of water.

To simplify the argument, let us suppose that the capillary tube has a uniform bore; the argument can be extended to cover all cases, but it then becomes rather long.

Suppose that the water-level inside the tube at the instant under consideration is at X, h cm. below C. Denote by α the temperature coefficient of linear expansion of glass and by β the temperature coefficient of cubical expansion of water.

Then $$r_1^4 = r_0^4\{1 + \alpha(\theta_1 - \theta_0)\}^4, \quad l_1 = l_0\{1 + \alpha(\theta_1 - \theta_0)\},$$

and since V is the volume of the tube between A and B,

$$V_1 = V_0\{1 + \alpha(\theta_1 - \theta_0)\}^3.$$

The excess pressure p is caused by the head h, so $p=h\rho g$ and

$$p_1=h_1\rho_1 g=h_0\{1+\alpha(\theta_1-\theta_0)\}\rho_0\{1-\beta(\theta_1-\theta_0)\}g$$
$$=p_0\{1-(\beta-\alpha)(\theta_1-\theta_0)\} \text{ nearly.}$$

Hence substituting these values and cancelling $\{1+\alpha(\theta_1-\theta_0)\}^4$,

$$\eta_1=\frac{\pi p_0 r_0^4 t_1}{8l_0V_0}\{1-(\beta-\alpha)(\theta_1-\theta_0)\}$$

and
$$\frac{\eta_1}{\eta_0}=\frac{t_1}{t_0}\{1-(\beta-\alpha)(\theta_1-\theta_0)\} \text{ nearly.}$$

Suppose that the object of the experiment was to determine the viscosity η_{30} of water at 30° C., being given that the viscosity at 20° C. is 0·00893 g. cm.$^{-1}$ sec.$^{-1}$, that α is about 8×10^{-6} and β is about $2{\cdot}60\times10^{-4}$. If we disregard the expansions of glass and water we should take η_{30} as being $0{\cdot}00893\times y_{30}/y_{20}$, where y_{20} and y_{30} are ordinates on the graph we obtained in the experiment. This value of η_{30} will be found to be roughly 0·00800.

But the foregoing theory shows that this last number should have been multiplied by the factor 1·0 – (0·000260 – 0·000008) (30 – 20), or

$$1{\cdot}0-0{\cdot}00252.$$

Hence our value of η_{30} will be about $0{\cdot}00252\times0{\cdot}0080$ too large, which means that we shall have made a percentage error of

$$\frac{0{\cdot}00252\times0{\cdot}0080}{0{\cdot}0080}\times100 \quad\text{or}\quad 0{\cdot}25\%.$$

This is much less than the probable experimental error, and can be ignored.

Exp. 35. *Cooling graphs*

The purpose of this experiment is to try out the merits of a method of reducing observations on a cooling body, which depends on mathematical theory, and to compare it with the more usual method of reduction, of drawing tangents to the cooling curve.

Take a small calorimeter and thoroughly smoke its outer surface; put in it about 50 c.c. of linseed oil, or turpentine; water will serve, but the experiment will take about double as long with water.

Heat the calorimeter and oil on a sand-bath (for safety) to about 80° C., put in it a stirrer and thermometer (preferably through two holes in a cardboard lid) and stand it either on a bench, or on a cork on the open bench, or in an outer weighted container standin in a pan of cold water, as for the 'Method of Cooling'.

With a watch or clock with a second hand note the time of passing each degree graduation until it has cooled to 40° C. or less, stirring steadily.

Complete the table of observations by a column showing for each degree-mark the number (t) of seconds elapsed after passing the first mark, another column showing (θ) the difference between the temperatures of the oil and the surrounding air, and another column showing $\log \theta$, in each case.

Plot a graph connecting $\log \theta$ and t.

You will probably find that a considerable part at least of this graph is a straight line, except for errors of observation. If so, denote by $-m$ the slope of the line to the axis of t, and by b its intercept on the axis of $\log \theta$.

Then the relation between θ and t, for a considerable range of θ, must be $\log \theta = b - mt$.

It is shown theoretically in the note appended to Exp. 16 that, for the range of θ over which a relation of this kind holds good, the time-rate of change of θ (i.e. the rate of cooling of the body) is proportional to θ, the difference between the temperature of the body and the temperature of its surroundings at any instant. Hence if your graph is a straight line you will have verified Newton's law of cooling.

Another way of reaching this result from the observations is to plot the 'cooling curve' connecting θ and t and draw tangents to it at various points, and show that the slopes of these tangents to the axis of t are roughly proportional to the values of θ corresponding to the points of contact of the tangents.

It is easy to see that the slope of the tangent is the rate of cooling, so the latter procedure is direct and straightforward, and well suited to beginners, by whom it is usually carried out. But it requires considerable skill to draw tangents to a curve with any accuracy and the method seldom gives convincing results. On the other hand, a satisfactory degree of accuracy can easily be attained with the former method, though some mathematics is needed for the deduction of the required relation from the straight-line graph. In fact, if the experiment is done twice, with the same calorimeter and oil, first on the open bench and then in a constant temperature enclosure, it is accurate enough to show a difference in the laws of cooling under these different conditions.

Exp. 36. *Thermal capacity of a vacuum flask*

There are many experiments in which a vacuum flask forms the most suitable calorimeter; the rate of loss of heat from its inner container is so small that the hot body can be stored there while bodies at about atmospheric temperatures are being transferred to the calorimeter; many of the usual experimental precautions to prevent unknown losses of heat are therefore unnecessary.

But its water equivalent is large enough to call for accurate measurement; this can be done experimentally as follows.

The error caused by assuming that the mass of 1 c.c. of water is 1 g. is negligibly small in this experiment since measurements of mass or volume are made at atmospheric temperatures; the volumes are best measured by a burette, delivering into a beaker.

Measure roughly the capacity (V_0 c.c.) of the flask; measure out accurately V_1 c.c. (rather less than half V_0) of cold water into a beaker, heat it to about 80° C., put it in the flask and cork the flask. Tip the flask about to ensure that the whole of the inner container and its contents are at the same temperature, measure this (θ_1° C.), estimating to 0·1° C. and keeping the flask uncorked for as short a time as possible.

Measure out V_2 c.c. (rather less than half V_0) of cold water into a beaker, take its temperature (θ_2° C.), pour it into the flask and take the final temperature (θ_3° C.) of the mixture, as before.

Calculate the thermal capacity (v g.) of the flask from the equation

$$(V_1+v)(\theta_1-\theta_3)=V_2(\theta_3-\theta_2),$$

or

$$v=V_2\frac{\theta_3-\theta_2}{\theta_1-\theta_3}-V_1.$$

Make several determinations of v and calculate its mean value; the time saved by measuring volumes instead of weights can well be devoted to repeating the observations with different temperatures and quantities of material. The mean difference (without regard to sign) between individual values of v and the mean value of v should not exceed 0·5 g.

The value of the thermal capacity so found should be treated as an instrumental constant of the particular flask, and marked on it for future use.

The number of Joules equivalent to one Calorie can be found with a very fair degree of accuracy if a vacuum flask with known thermal capacity is used as a calorimeter. A loosely wound coil of resistance wire of known resistance, soldered to thick copper leads, is put in a measured mass of cold water in the flask, and a steady current is passed through the coil; the P.D. between the ends of the coil is measured by a voltmeter, so that the number of watts dissipated in the coil can be calculated. The rise in temperature of the water in the flask in a measured time is observed. It is left as an exercise for the resourceful student to design the details of this measurement, and to carry it out.

Exp. 37. *Newton's law of cooling*

When a current flows through a bare wire, hung in air, the temperature of the wire rises and its length increases. This is the principle underlying the Hot Wire Ammeter; in this experiment it is used to provide an indirect method of testing the truth of Newton's law of cooling.

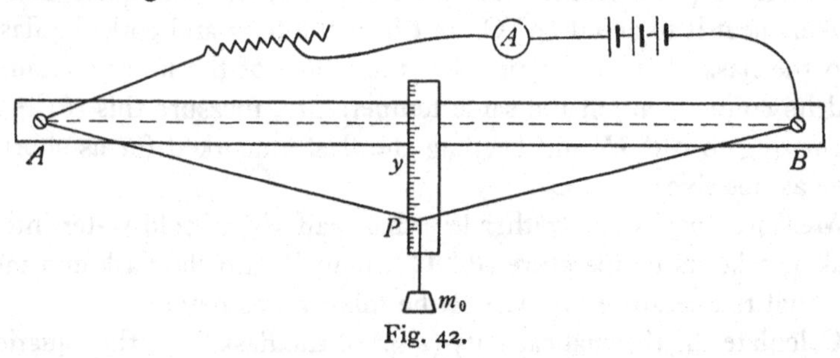

Fig. 42.

Screw two suitable binding screws A and B in the edge of a board upwards of a metre long, fixed so that its faces are horizontal, and stretch between A and B a piece of bare resistance wire, of negligible temperature coefficient, and of about No. 32 S.W.G., with a tension sufficient to make the wire nearly straight but not sufficient to give an audible fundamental tone when plucked. Make certain that the wire cannot slip at A or B.

Put in series with the wire a variable rheostat, a battery of three secondary cells and an ammeter A reading to 3·0, or at least to 1·5, amps.

Hang from the centre P of the wire a 50 g. weight, and fix a vertical scale so that the distance (y) of P below AB can be measured.

Pass through the wire a succession of currents of observed amounts, i amps., rising to the maximum available and then decreasing to zero; observe in each case the corresponding value of y (including that for $i=0$, denoted by y_0). Calculate i^2 and $y^2-y_0^3/y$ in each case, and plot a graph with $y^2-y_0^3/y$ as abscissae and i^2 as ordinates.

The appended note on the theory shows that when y is small, the difference of temperature between the wire and its surroundings is equal to $\frac{2}{AB^2\alpha}\left(y^2-\frac{y_0^3}{y}\right)$, where α is the coefficient of thermal expansion of the wire; if Eureka wire is used, α may be taken as $1\cdot7\times10^{-5}$. Hence the abscissae of the graph are proportional to this temperature difference. Now i^2 is proportional to the rate at which heat is being generated in the wire by the current, since its resistance is independent of its temperature; since the temperature of the wire is steady, i^2 is therefore proportional to the rate at which heat is being dissipated to the surroundings of the wire. Hence the ordinates of the graph are proportional to the rate at which the wire would cool if the current did not prevent it doing so. Therefore, in so far as Newton's law of cooling holds good, the graph should be a straight line.

This law is true only for a limited range of temperature differences; so your graph may be straight at first but become curved beyond a certain point.

Note on the Theory

Suppose that a wire is attached to two points in a horizontal line $2L$ apart, the wire being practically straight but with negligibly small tension in it. Suppose that a small mass, m, is hung at its centre P, and that P then drops to P_0 through a distance y_0 below AB, and that P_0B is then equal to x_0 and that the tension in it is then T_0.

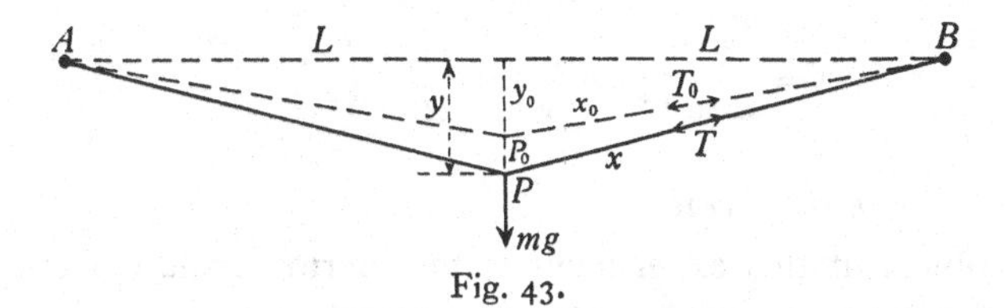

Fig. 43.

Then

$$x_0^2=L^2+y_0^2, \quad \text{or} \quad x_0=L\left(1+\frac{y_0^2}{L^2}\right)^{\frac{1}{2}}=L\left(1+\frac{y_0^2}{2L^2}\right) \text{ nearly}$$

$$=L+\frac{y_0^2}{2L} \text{ nearly.}$$

Hence the extension of BP due to the weight is $y_0^2/2L$.

But this extension is $\frac{L}{EA}T_0$, where E is Young's modulus and A is the area of cross-section of the wire. So

$$\frac{LT_0}{EA}=\frac{y_0^2}{2L}. \tag{1}$$

But $$mg=2T_0\frac{y_0}{x_0}, \quad \text{or} \quad T_0=\frac{mgx_0}{2yo},$$

and from (1) $$\frac{mg}{2EA}=\frac{y_0^3}{2L^2x_0}. \tag{2}$$

Now let the temperature of the wire be raised by any means through $\theta°$ C.; denote by α its coefficient of linear thermal expansion. Then the extension of BP due to this rise of temperature is $L\alpha\theta$. Suppose that P drops to a total distance y below AB in consequence of this thermal expansion and the tension T of the wire caused by the weight m, and that $BP=x$.

Then, as above, total extension $=\frac{y^2}{2L}$ and $mg=2T\frac{y}{x}$. (3)

This total extension is the sum of the thermal expansion $L\alpha\theta$ and the elastic extension $\frac{L}{EA}T$ or $\frac{Lmgx}{2EAy}$ from (3), so

$$\frac{y^2}{2L}=L\alpha\theta+\frac{Lmgx}{2EAy}=L\alpha\theta+\frac{Lxy_0^3}{2L^2x_0y} \quad \text{from (2).}$$

Since x_0 and x are nearly equal to L, this can be written as

$$\frac{y^2}{2L}=L\alpha\theta+\frac{y_0^3}{2Ly},$$

or $$\theta=\frac{1}{2L^2\alpha}\left(y^2-\frac{y_0^3}{y}\right)=\frac{2}{AB^2\alpha}\left(y^2-\frac{y_0^3}{y}\right).$$

Exp. 38. *Immersion heater*

The object of this experiment is to determine the current to be supplied to an immersion heater in a vessel containing warm water in order to maintain it at a constant given temperature in spite of the loss of heat to its surroundings, and to check the accuracy of that determination.

Let us first determine the number of calories to be supplied per second to the given vessel of hot liquid standing on the open bench, in order to keep its temperature steady.

Take, for example, a small calorimeter, of measured mass m_1 g. and specific heat s_1, containing a measured mass m_2 g. (say about 100 g. if that nearly fills the calorimeter) of a liquid of specific gravity ρ_2 and specific heat s_2; denote by M g. the thermal capacity of the whole; we can ignore the thermal capacity of the thermometer to be used. Then $M = m_1 s_1 + m_2 \rho_2 s_2$. It saves trouble in our experiment to use water as the liquid.

Suppose that the required steady temperature is 45° C. Heat the calorimeter and water to about 48° C., stand it on a cork on the bench, stir steadily but gently, and note the times when its temperature is 46, 45 and 44° C. Hence calculate the number n of seconds taken to drop 1° C. when the temperature is at the mid-point, 45° C.

For a drop of 1° C., M calories have been lost by the vessel, so the number of calories to be supplied per sec. to maintain the temperature steady is M/n. This is equivalent to $4{\cdot}2M/n$ watts.

A convenient immersion heater for this experiment consists of a length of insulated resistance wire, soldered to fairly thick copper leads, with the resistance wire coiled tightly round the lower end of a narrow test-tube, and the leads tied to opposite sides of its upper part. It will be seen that, since the temperature of the heater is to remain steady while it is in use, its thermal capacity need not be known. A convenient resistance is about 2 ohms; this can be got by using about 70 cm. of 26 S.W.G. Eureka or constantan wire. It is a refinement, but in this experiment an almost unnecessary refinement, to provide a separate pair of leads from the ends of the resistance wire, which can be connected to the voltmeter used for measuring the P.D. across the resistance wire; in practice we can with sufficient accuracy attach this voltmeter to the outer ends of the current-carrying leads.

The resistance, R ohms, of the heater must first be determined. This can, of course, be done by a Wheatstone's bridge method, but it will be sufficiently accurate here to measure with the voltmeter the P.D. (V_1 volts) across it when a current (C_1) of about an ampere is passed through it and an ammeter by a single cell; then $R = V_1/C_1$.

Then if a sufficient current, C amps., is passed to maintain a P.D. of V volts across it and to produce an expenditure of $4{\cdot}2M/n$ watts in it, we must have $\frac{4{\cdot}2M}{n} = V \times C = V \times \frac{V}{R} = \frac{V^2}{R}$.

Hence we can calculate V, the P.D. which must be maintained across the terminals of the heater in order to maintain the steady temperature of the liquid in which it is immersed.

The value of V should be calculated for the above conditions from the measured values of M, n and R, and the accuracy of the theory (including Joule's Law) on which that calculation was based can be checked by the following experiment.

Heat the calorimeter and water again to about 48° C., put in the immersion heater and thermometer, connect a voltmeter across the heater terminals in parallel with a 6-volt direct current supply (such as three accumulators) and a variable rheostat. Adjust the rheostat until a P.D. of V volts (as calculated above) is shown; switch off the current and let the liquid cool until it reaches 45° C. when stirred.

At this point switch on the current, stir steadily and whenever necessary adjust the rheostat to keep the voltage constant, and see whether the temperature remains steady at 45° C. for five or ten minutes (as it should).

Exp. 39. *Change of resistance with temperature*

The object of this experiment is the investigation of the relation between the resistance of a copper wire and its temperature; in order that it may be carried out with comparatively small currents the copper wire should be thin, not more than No. 36 S.W.G.

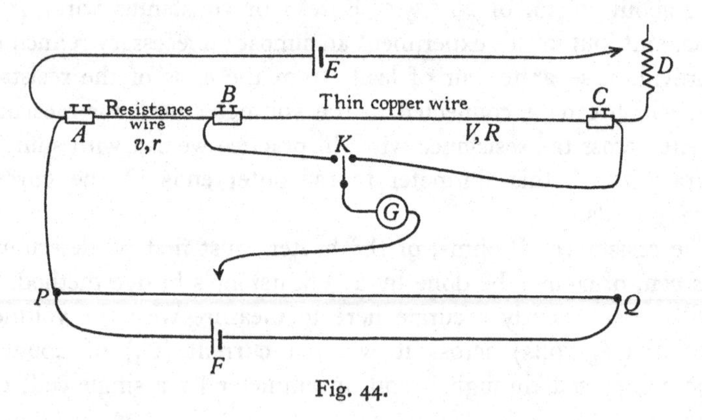

Fig. 44.

Take a length BC (about 50 cm.) of this (uncovered) copper wire; it will be convenient for setting up (but not essential) if its ends are soldered to two short lengths (say 1 cm. each) of thick copper wire. Fix these thick wires in connectors B and C. In B fix one end of a length *BA* of resistance wire of negligible temperature coefficient, which has roughly the same resistance (about 0·3 ohm) as the copper wire; for instance, about 25 cm. of No. 22 or 15 cm. of No. 24, or two lengths each of 20 cm. in parallel of No. 26 S.W.G., Eureka wire; the other end of this resistance wire is to be connected by a connector *A*, as in the figure, to the end *P* of a potentiometer wire *PQ*, and by a fairly thick copper wire to a single accumulator *E*.

Support *A* and *C* in retort stands so that *ABC* hangs clear of the bench.

D is a rheostat which can carry 2 or 3 amps., of maximum resistance about 3 ohms. *K* is a two-way key by which the galvanometer *G* of the potentiometer *PQ* can be connected to either *B* or *C*. The potentiometer has its own accumulator *F*, which we assume will maintain a constant current along *PQ*.

If *E* sends a current, *i* amp., through *ABC*, it will cause a P.D. (v volts, say) between *A* and *B*, and a P.D. (V volts, say) between *B* and *C*. Let *G* be connected to *B* and the null point on *PQ* be found to be l_1 cm. from *P*; then let *G* be connected to *C* and the null point be found to be l_2 cm. from *P*. Then $\frac{l_2}{l_1}=\frac{V+v}{v}$, so that $\frac{V}{v}=\frac{l_2-l_1}{l_1}$.

Denote the resistances of *BC* and *AB* by *R* and *r* ohms respectively; then $V=Ri$ and $v=ri$, so that $\frac{V}{v}=\frac{R}{r}$, and finally $R=r\frac{l_2-l_1}{l_1}$.

Now both *AB* and *BC* will be to some extent heated by the current; this will not change *r*, but it may change *R*, and if it does the change will be shown by a change in $\frac{l_2-l_1}{l_1}$ when the current is changed.

In the appended note it is shown theoretically that, in so far as Newton's law of cooling holds good, the difference of temperature (t° C.) between the copper wire *BC* and its surroundings is directly proportional to Vv or $(l_2-l_1)\,l_1$.

If then we measure l_1 and l_2 for a series of settings of the rheostat, from zero resistance to its maximum, and plot the graph connecting $\frac{l_2-l_1}{l_1}$ and $(l_2-l_1)\,l_1$, the graph will show the form of the relationship between *R* and *t*.

If you find, as is probable, that this graph, or part of it, is a straight line not passing through the origin, the corresponding relation between *R* and *t* will be linear, of the form $R=A+Bt$, where *A* and *B* are constants; if its intercept on the R axis is R_0, we can write it in its familiar form $R = R_0(1 + at)$, where a is a constant.

You will probably find that the dimensions of the apparatus given above enable you to get a substantial change of R, and that when the current is at its maximum the copper wire feels quite hot, so that although there is here no method of measuring t it may obviously be considerable. Since the graph is linear only so far as Newton's Law holds good, the law must hold good over a considerable range of temperature-difference.

Note on the Theory

If I amps. is the current through PQ and ρ ohms is the resistance per cm. of PQ, then the P.D. between A and B (which we have denoted by v volts) is $l_1 \rho I$, and the P.D. between B and C (or V volts) is $(l_2 - l_1)\rho I$. So $v = l_1 \rho I$ and $V = (l_2 - l_1)\rho I$.

Now by Joule's Law the heat generated per sec. in BC is 0·24 Vi calories, or

$$0{\cdot}24 V \frac{v}{r} \quad \text{or} \quad 0{\cdot}24(l_2 - l_1)\frac{\rho I l_1 \rho I}{r} \quad \text{or} \quad \frac{0{\cdot}24\rho^2 I^2}{r}(l_2 - l_1) l_1 \text{ calories.}$$

Hence if the current I is constant, the heat generated per sec. in BC is directly proportional to $(l_2 - l_1) l_1$; since the temperature of BC soon settles down to a constant value, this heat must be dissipated at the same rate to the surroundings of the wire. Now by Newton's Law of cooling, this rate of dissipation is proportional to the difference of temperature (t° C.) between BC and its surroundings, so t is directly proportional to $(l_2 - l_1) l_1$ for the range of t for which that law holds good.

Exp. 40. *Resistance of a coil carrying a current*

If we want to measure the resistance of a wire we must pass a current through it, in order to bring into play the resistance to be measured; this current will heat the wire to some extent; hence our experiment will give the resistance at some temperature, t° C. say, above that of the air, and it is not usually practicable to measure t directly.

This does not matter if the wire has a zero temperature coefficient, as is very nearly the case with such alloys as manganin, Eureka, constantan, etc., of which standard resistance coils are made, but it does matter with a copper wire, of which the coils in instruments, dynamos, etc., are made.

If we are using a Wheatstone's bridge, and cut down the current until it is almost zero, in the hope of measuring the resistance of the coil at almost exactly air temperature, we shall at the same time proportionately reduce the P.D. indicated by the galvanometer when connected to the bridge wire at a point other than the balancing point; since there is a limit to the sensitiveness of the galvanometer, we should then be unable to find the balancing point.

The following experiment shows how this difficulty can be overcome without any such reduction of the current in use, and the resistance of the coil when carrying no current at all can be determined indirectly.

Take a length of about 150 cm. of No. 36 S.W.G. insulated copper wire and wind it smoothly round a match stick, whose corners have been smoothed off so that it is roughly circular, with consecutive turns of the coil in close contact like the thread round the handle of a cricket bat. You should begin by laying the wire along the match, so that it projects about 7 cm. beyond one end of the match, and start winding over this wire at the other end of the match, continuing until you have put on 50 cm., making a coil about 1·5 cm. long. Then wind backwards on top of the first layer until you reach its other end; then wind on a third layer until you come to the end of the coil. Twist together the two loose ends of wire, for two or three turns, where they leave this point. Keep the three-layered coil thus formed on the match and connect it in one of the gaps of a metre wire bridge. A 50 cm. bridge will serve if no metre bridge is available, but even more care will then have to be taken in the accurate location of the balancing point; this must in any case be estimated to tenths of a millimetre.

Fig. 45 shows X, the coil whose resistance is required, in the left-hand gap of a wire bridge, and a 1-ohm coil in parallel with a high-resistance voltmeter, reading from 0 to 1·0 volt, in the right-hand gap. This coil will not be required to carry more than half an ampere, so the type of standard coil usually used in elementary laboratories can safely be used here.

The battery for the bridge should consist of one or two accumulators in series, and a rheostat, which does not heat up unduly with half an ampere, should be put in series with it. The readings of the voltmeter, which are nominally volts, will represent the number of amperes flowing through the 1-ohm coil, and through the coil X provided that P is the balancing point so that no current is diverted through G the galvanometer. Adjust the rheostat until the voltmeter reads exactly 0·2 volt, so that i, the current through X, is 0·2 amp.; find the balancing point after the current has passed long enough for the steady state to be attained; this will require a considerable time. The position of P should be found at intervals of one or two minutes until there is no perceptible change between readings two minutes apart. The reading of V must remain constant, or be kept constant by adjusting the rheostat.

Calculate BP/AP, the reciprocal of R the resistance of X; you should use four-figure logs.

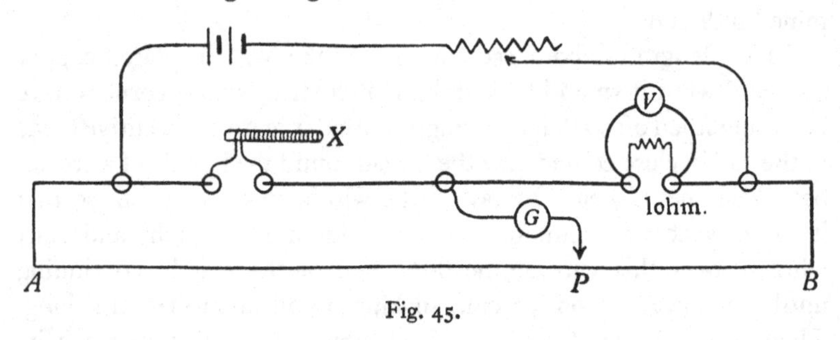

Fig. 45.

Then adjust the rheostat so that the value of i is exactly 0·3, 0·4 and 0·5 amp. in succession and find the corresponding values of $1/R$.

Plot a graph connecting $1/R$ as ordinates with i^2 as abscissae. It will probably be nearly a straight line; if so, find the intercept of the best-fit straight line on the axis of $1/R$.

There is every reason to suppose that this intercept will be the value of the reciprocal of R_0, the resistance of the coil when it carries zero current. Calculate R_0 on this assumption; it is what we set out to find.

Other experiments show that the relation between the resistance R of a copper wire and its temperature is very nearly $R=R_0(1+at)$, where $a=0{\cdot}00428$, R_0 is the resistance of the wire at air temperature and t is the difference between the temperatures of the wire and the air. Hence $t=\dfrac{R-R_0}{0{\cdot}00428R_0}$.

You have calculated the value of $1/R$ for each of your observations; find by means of a table of reciprocals the corresponding values of R, and hence the values of t from the above relation. Plot a graph connecting t and i^2. This will probably turn out to be a straight line through the origin.

If the insulation on the wire is thick enough to offer an appreciable resistance to the flow of heat from one layer of the coil to the next layer, the innermost layer will be at a higher temperature than the next layer, and so on, when the steady state is attained. Hence the resistance of the innermost layer will be higher than the resistance of the next layer, and so on, and R will be the sum of a number of different resistances, while t will be a kind of average temperature of the coil as a whole. So the innermost layer may be a good deal hotter for a given current than the temperature shown on your third graph, and may even be dangerously high while the outside layer is not unduly hot; this must be taken into account in estimating the safe current for a given coil.

The usual 'wiring rules' lay down 500 amps. per sq.cm. as a safe current density for wires thinner than N° 22 S.W.G., which gives a safe current of 0·15 amp. for N° 36 wire. You should determine from your second graph the corresponding value of t for your coil; you will probably find that in your experiment you have greatly exceeded this safe current, without ill effects; but if the wire had been made of constantan instead of copper, t would have been about 25 times as large, for the same currents.

Note on the Theory

Assume for simplicity that the insulation of the copper wire in use in this experiment is so thin that it does not appreciably affect the distribution of temperature in the coil when it attains the steady state. Then, in this state, the temperature of the whole of the wire will be the same, and, say t° C. above air temperature. Denote by R_0 ohms the resistance of the coil when no current is running, and by R ohms its resistance when a current i amps. has been running long enough for the steady state to be attained.

Then $R = R_0(1 + \alpha t)$, where α is a constant, 0·00428 for copper.

By Joule's Law, $\dfrac{i^2R}{4{\cdot}2}$ calories are generated in each second; since we have a steady state of things, this heat flows mainly through the exposed surface of the coil, but partly through the prolongation of the core. In either case, if Newton's law of cooling holds good, $\dfrac{i^2R}{4{\cdot}2} = At$, where A is a constant.

But $$t = \frac{R - R_0}{\alpha R_0}, \quad \text{so} \quad \frac{i^2R}{4{\cdot}2} = A\,\frac{R - R_0}{\alpha R_0}.$$

Rearranging this, $$\frac{1}{R}+\frac{\alpha}{4\cdot 2A}i^2=\frac{1}{R_0}, \qquad (1)$$

Since the coefficient of i^2 is a constant, the graph connecting i^2 and $1/R$ is a straight line, whose intercept on the axis of $1/R$ is $1/R_0$.

Substituting $R_0(1+\alpha t)$ for R in (1), if t is only moderately large so that αt is small compared with 1·0 we get

$$\frac{\alpha}{4\cdot 2A}i^2=\frac{1}{R_0}-\frac{1}{R_0}(1+\alpha t)^{-1}=\frac{\alpha t}{R_0}\text{ nearly,}$$

or $$t=\frac{R_0}{4\cdot 2A}i^2\text{ nearly.} \qquad (2)$$

Hence, in a given coil of this kind, the rise of temperature above that of the air is directly proportional to the square of the current.

It can be shown that, if the resistance of the insulation to the flow of heat is not negligibly small, the above results hold good precisely when the coil consists of a single layer, except that the coefficient of i^2 in (2) is a different constant, involving the heat-conductivity of the insulation as well as the rate of emission from its surface; when it has several layers, they are very approximately true, as the experiment will most likely show. But the analysis in that case becomes rather too long to be worth setting out here.

Exp. 41. *Resistance of a voltmeter*

It is generally taken for granted that a voltmeter has a 'very large' internal resistance, so large, in fact, that no inquiry need be made into its actual value. But in some cases it is necessary to know that value, at any rate approximately, and the aim of this experiment is to determine it, by using the voltmeter itself as an ammeter, in conjunction with a resistance box which will give a total resistance of 500, or perhaps 1000, ohms.

Connect in series the voltmeter V whose resistance (X ohms) is required, the resistance box R and a battery of accumulators whose E.M.F. (E volts) should be rather larger than the maximum reading of the voltmeter. It will be assumed that the internal resistance of this battery is negligibly small compared with X.

Adjust R until the voltmeter reading is near the maximum of its scale, showing V volts. The ordinary resistance box enables us to vary R, when it is large, by steps that are small compared with R; we shall get greater accuracy in this experiment if we adjust R to bring the voltmeter needle as nearly as possible to one of the graduations of the scale, and thus reduce, or almost entirely eliminate, the uncertainty of estimating tenths of a division.

Increase R step by step until the voltmeter shows about half its maximum deflection; record corresponding values of R and V, and plot a graph connecting R and $1/V$. If as is probable it is a straight line, find the intercepts on both axes.

To make $R=0$ is obviously equivalent to connecting the voltmeter directly to the terminals of the battery, so that the intercept on the $1/V$ axis must equal $1/E$; E is then the P.D. between the terminals of the voltmeter, whether or not the actual voltmeter in use can read so large a voltage; hence the graph enables us to measure E in either case, as the reciprocal of this intercept. Hence determine E.

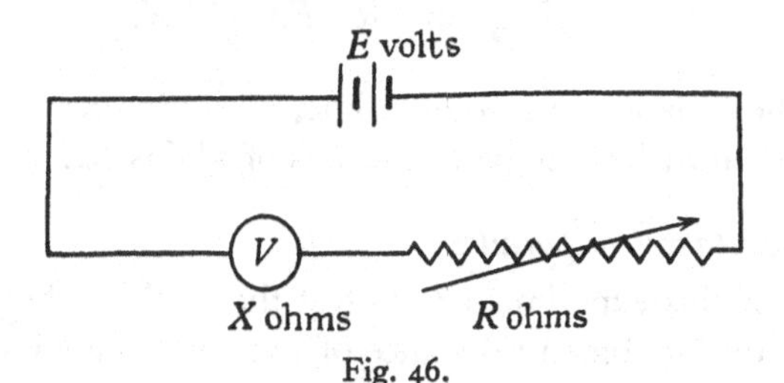

Fig. 46.

But the resistance box does not provide a resistance less than zero, that is a negative resistance, so the experiment cannot give us points on the left-hand side of the axis of $1/V$, which would correspond to negative values of R. Nevertheless, we can if we like produce geometrically the graph to intersect the axis of R and use mathematically the value of the intercept; mathematical equations hold good whether R is positive or negative. Hence, if we can get by theory a mathematical relation between X and this intercept, we can determine X by measuring the intercept. The appended note shows that it should theoretically be $-X$. Hence determine X.

The theory given in the note also shows that the slope of the graph to the axis of $1/V$ is EX; we have seen how we can determine E from the intercept on the axis of $1/V$, so we can determine X by measuring the slope of the graph, without producing it geometrically into the region where R is negative, but using only the part of the straight line covered by observations; we ought to get the same value for X whichever course we adopt; you should check this.

Note on the Theory

Denoting by i amps. the current in the circuit, Ohm's Law gives

$$i=\frac{E}{X+R}, \tag{1}$$

where R is the resistance brought into use in the resistance box.

Again, the voltmeter measures the P.D. (V volts) between its terminals. Since the current through it is i amps. by Ohm's Law,

$$i=\frac{V}{X}. \tag{2}$$

Hence by (1) and (2)

$$\frac{V}{X}=\frac{E}{X+R} \quad \text{or} \quad R=EX\frac{1}{V}-X. \tag{3}$$

This is the equation of a straight line, whose intercept on the axis of R is $-X$, and whose slope to the axis of $1/V$ is EX.

Exp. 42. *Resistances in parallel*

The aim of this experiment is to test the truth of the theoretical formula for the combined resistance of two resistances in parallel.

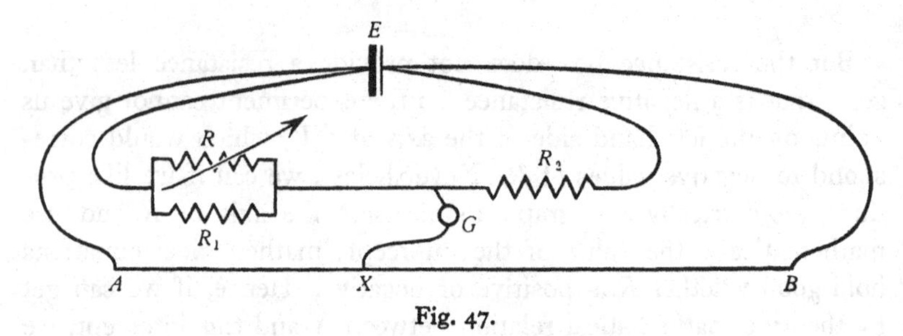

Fig. 47.

Connect up apparatus as in Fig. 47, where R_1 is a 10-ohm coil, R_2 a 5-ohm coil, R a resistance box whose largest resistance is not less than 5 ohms, AB a potentiometer wire of length L cm., and G a galvanometer. E is a cell, such as a Leclanche or an accumulator; there is no need for its resistance to be very low, or its E.M.F. constant during the experiment.

Make very sure that you have good contact at all terminals.

Measure the distance, x cm., from A of the balancing point X for various values of R; in order to get points well spaced along the graph, use all available small values of R and few large values.

Plot a graph with L/x as ordinate and $1/R$ as abscissa. If it is a straight line, find its intercept on each axis. The appended note shows that the relation between L/x and $1/R$ should theoretically be

$$\frac{L}{x}=R_2\frac{1}{R}+1+\frac{R_2}{R_1},$$

so your graph should theoretically be a straight line, and its intercept on the axis of L/x should be $1+R_2/R_1$, or $1\cdot 50$ with the above values of R_2 and R_1, and its intercept on the axis of $1/R$ should be $-\frac{1}{R_2}\left(1+\frac{R_2}{R_1}\right)$ or $-\left(\frac{1}{R_2}+\frac{1}{R_1}\right)$ or $-0\cdot 3$ in this case.

Compare your experimental results with these values. The theory in the appended note depends on the truth of the usual formula for the resistance of two resistances in parallel; if you confirm it, you indirectly confirm the truth of that formula.

Note on the Theory

Whatever be the E.M.F. or resistance of the battery, if there is no current through G,

$$\frac{\text{Resistance of } R \text{ and } R_1 \text{ in parallel}}{\text{Resistance of } R_2}=\frac{\text{P.D. across } R \text{ and } R_1 \text{ in parallel}}{\text{P.D. across } R_2}$$

$$=\frac{\text{P.D. across } AX}{\text{P.D. across } XB}=\frac{x}{L-x}.$$

Hence

$$\frac{\text{Reciprocal of resistance of } R \text{ and } R_1 \text{ in parallel}}{1/R_2}=\frac{L-x}{x}.$$

The formula for the numerator of the fraction on the left-hand side is $1/R+1/R_1$. Hence

$$\frac{\frac{1}{R}+\frac{1}{R_1}}{\frac{1}{R_2}}=\frac{L-x}{x}=\frac{L}{x}-1.$$

Therefore

$$\frac{L}{x}=1+R_2\frac{1}{R}+\frac{R_2}{R_1}, \quad \text{or} \quad \mathbf{\frac{L}{x}=R_2\frac{1}{R}+1+\frac{R_2}{R_1}}.$$

Exp. 43. *Errors due to inadequate voltmeter resistance*

The aim of this experiment is to investigate the errors caused by inadequate resistance in a voltmeter when it is used to measure the P.D. across a resistance in a circuit carrying a current.

Set up the circuit shown in the figure, in which the voltmeter (V) has a known resistance R_1 ohms, the battery has an E.M.F. (E volts) which should be between once and twice the maximum reading of the voltmeter (which we will assume to be about 1·5 or 3·0 volts). The resistance between B and C is a resistance coil of known value (R_2 ohms) which we will take to be 20 ohms, and the variable resistance between A and B is a resistance box giving a total of about 50 or 100 ohms by single ohms.

Take out all the plugs in the resistance box; reduce its resistance until the voltmeter records the voltage (V) between A and B. Then reduce R step by step, recording the corresponding value of V. As explained in Exp. 41, it is advisable to adjust R in each case so that the needle of the voltmeter points as nearly as possible to one of its scale divisions.

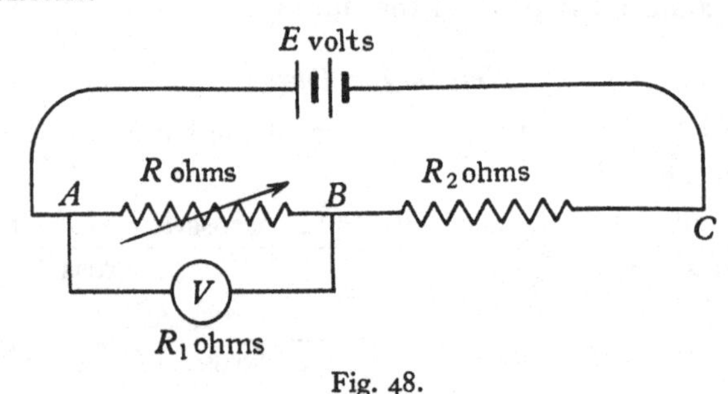

Fig. 48.

Plot a graph connecting $1/V$ with $1/R$. If this turns out to be a straight line, determine its slope to the axis of $1/R$ and its intercept on the axis of $1/V$.

Determine the value of E; if the battery consists of more than one cell and the voltmeter reads up to 2·0 volts you can do this by measuring the E.M.F. of each cell separately with that voltmeter; otherwise, use another voltmeter with a larger range.

The appended note shows that the theoretical relation between V and R is

$$\frac{1}{V}=\frac{R_2}{E}\frac{1}{R}+\frac{R_2}{E}\left(\frac{1}{R_1}+\frac{1}{R_2}\right), \tag{1}$$

so that the above slope should be R_2/E and the above intercept should be $\frac{R_2}{E}\left(\frac{1}{R_1}+\frac{1}{R_2}\right)$.

Substitute in these expressions the known values of E, R_1 and R_2 and compare the results with the values obtained from your graph.

If the results are concordant enough to verify (1), you are justified in adopting the mathematical consequences worked out in the note, that the voltmeter reading is $\dfrac{100 \times \dfrac{1}{R_1}}{\dfrac{1}{R}+\dfrac{1}{R_1}+\dfrac{1}{R_2}}$ % lower than the voltage between A and B when the voltmeter is not connected to these points; that is, that this is the effective percentage error of the voltmeter when used in this way, apart from any errors in its graduation, etc.

But if you examine your procedure critically you will probably find that you have not really confirmed the truth of (1), and if you have not done so you are not justified in drawing from your experiment the foregoing conclusion about the voltmeter error.

Voltmeters usually to be found in laboratories have a resistance of 200 to 300 ohms per volt of the maximum reading, so this voltmeter may have a resistance of from 300 to 900 ohms. Hence the above intercept may have a value (since we took $R_2 = 20$) of perhaps $20(\frac{1}{900}+\frac{1}{20})/E$. It is at least improbable that your measurement of the intercept of the graph will have an accuracy sufficient to make it safe to assert that the first term within the bracket is $\frac{1}{900}$ and not $\frac{2}{900}$, though it may be sufficient to verify the value $\frac{1}{20}$ for the second term. And it is on the magnitude of the first term that the final percentage error will be seen largely to depend, since the numerator of the fraction is simply 100 times this term.

The instructions were as a matter of fact framed to produce this result and to illustrate the need of vigilance. If you had originally taken the fixed resistance R_2 as 200 or 400 instead of 20 ohms, and a resistance box capable of making R about 500 ohms, your experimental results might have justified the theory to a reasonably satisfactory extent, since the two terms within the bracket would then have been comparable.

The matter is so important that it is worth while to repeat the experiment under these conditions, and to add to each of the graphs obtained the theoretical straight lines that would have been got if R_1 was infinitely large, that is, if the voltmeter had not been in place. The equation of these latter lines is, by (1), $\dfrac{1}{V}=\dfrac{R_2}{E}\left(\dfrac{1}{R}+\dfrac{1}{R_2}\right)$.

Your graphs will probably resemble these figures, in which Fig. 49 *a* shows the graphs calculated for the conditions $R_1 = 600$ ohms,

$E=4$ volts and $R_2=200$ ohms, and Fig. 49*b* is for the conditions $R_1=600$ ohms, $E=4$ volts, but $R_2=20$ ohms. The voltmeter is supposed to read up to 3 volts.

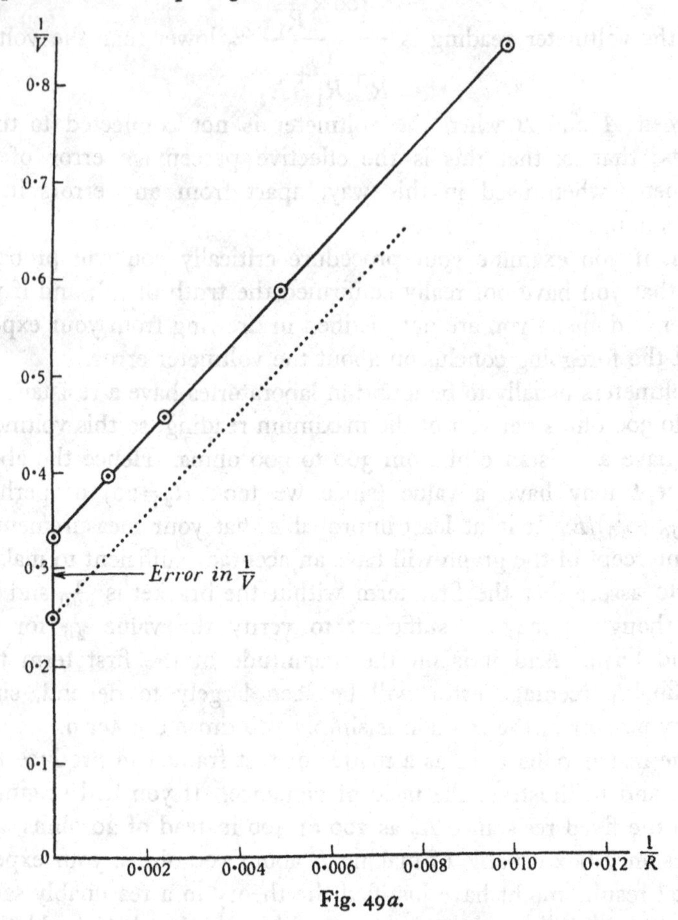

Fig. 49*a*.

In Fig. 49*a* the difference between these two graphs is substantial, and not likely to be seriously affected by casual experimental errors, when they have been smoothed out by the process of drawing the full line graph. Looking at this point in another way, the following table gives a few (calculated) values for this set-up.

R ohms	$1/R$	$1/V$	V volts	V_0 volts
500	0·0020	0·433	2·31	2·86
400	0·0025	0·458	2·18	2·67
200	0·0050	0·583	1·72	2·00
100	0·0100	0·833	1·20	1·33

The last column shows the voltage between A and B when there is no voltmeter between them, and the difference between this and the last column but one is the error due to inadequate voltmeter resistance, which is quite readable for the larger values of R.

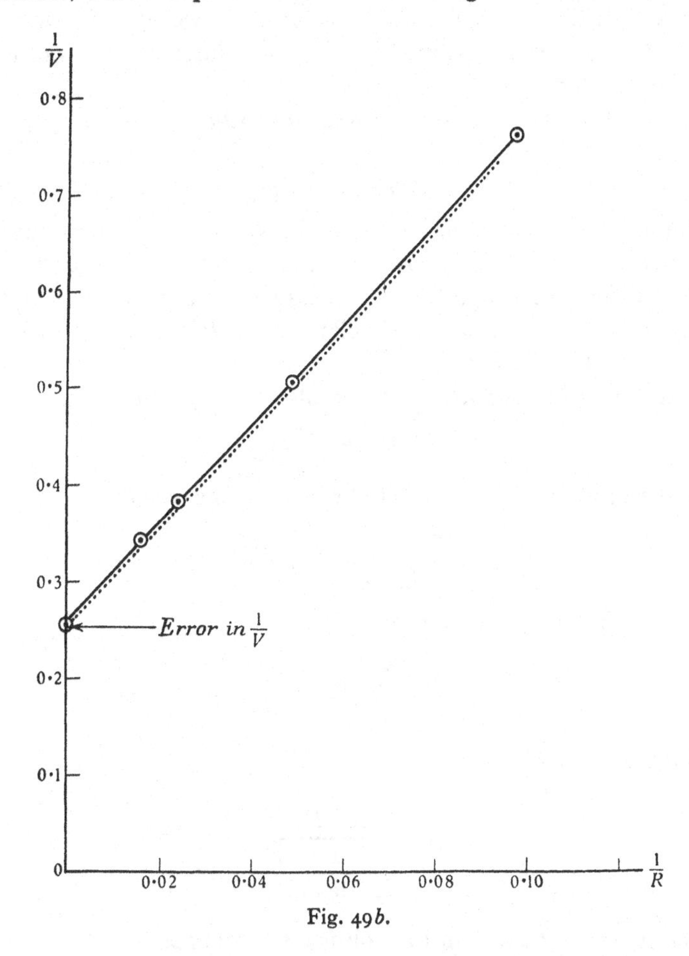

Fig. 49*b*.

Contrast with this Fig. 49*b* and the following table, which are calculated for the conditions and instructions for the first part of the experiment, $E=4$ volts, $R_1=600$ ohms and $R_2=20$ ohms.

R ohms	$1/R$	$1/V$	V volts	V_0 volts
50	0·020	0·358	2·79	2·86
40	0·025	0·383	2·61	2·67
20	0·050	0·508	1·97	2·00
10	0·100	0·758	1·32	1·33

The difference between the full and dotted graphs in Fig. 49*b* is so small that experimental errors would probably mask it, unless the procedure of Introduction (5) was adopted for drawing the best-fit line; and the second table shows that hundredths of a volt must be estimated with a high degree of accuracy if we are to bring out accurate values of the errors, which are the differences between the last two columns.

Hence R_2 must be large if we want to verify (1).

Note on the Theory

If a current of i_1 amps. flows through R in the circuit of Fig. 48 and a current of i_2 amps. flows through V, the current through R_2 and E will be i_1+i_2 amps. By Ohm's Law the P.D. between A and B, recorded by the voltmeter as V volts, will be Ri_1, and it will also be R_1i_2; hence $V=Ri_1=R_1i_2$.

And the P.D. between A and B is also $E-R_2(i_1+i_2)$. So

$$V=E-R_2(i_1+i_2).$$

Eliminating i_1 and i_2 from these three equations we get

$$\frac{1}{V}=\frac{R_2}{E}\left(\frac{1}{R}+\frac{1}{R_1}+\frac{1}{R_2}\right). \tag{1}$$

Denote by V_0 the value of V if R_1 is infinitely large, then

$$\frac{1}{V_0}=\frac{R_2}{E}\left(\frac{1}{R}+\frac{1}{R_2}\right). \tag{2}$$

Dividing (2) by (1) we get

$$\frac{V}{V_0}=\frac{\frac{1}{R}+\frac{1}{R_2}}{\frac{1}{R}+\frac{1}{R_1}+\frac{1}{R_2}}. \tag{3}$$

The percentage error in the voltmeter reading is

$$\frac{V-V_0}{V_0}\times 100, \quad \text{or} \quad 100\left(\frac{V}{V_0}-1\right), \quad \text{or by (3)} \quad \frac{-100\frac{1}{R_1}}{\frac{1}{R}+\frac{1}{R_1}+\frac{1}{R_2}}. \tag{4}$$

Writing (1) in the form

$$\frac{1}{V}=\frac{R_2}{E}\frac{1}{R}+\frac{1}{E}+\frac{R_2}{E}\frac{1}{R_1},$$

we see that it is the equation of a straight line inclined to the axis of $\frac{1}{R}$ at a slope of $\frac{R_2}{E}$ and making an intercept on the axis of $\frac{1}{V}$ of $\frac{1}{E}+\frac{R_2}{E}\frac{1}{R_1}$. So the slope is independent of R_1, and the error in $\frac{1}{V}$ produced by R_1 is R_2/ER_1; if E and R_1 are kept constant the error is proportional to R_2, as shown in Figs. 49*a* and 49*b*.

Exp. 44. *Resistance of a stretched wire*

If a copper wire is stretched beyond its elastic limit, it takes up a permanent set; the following experiment is designed to investigate the changes, if any, in the specific resistance of the copper caused by this extension. There may also be changes in the volume of the copper.

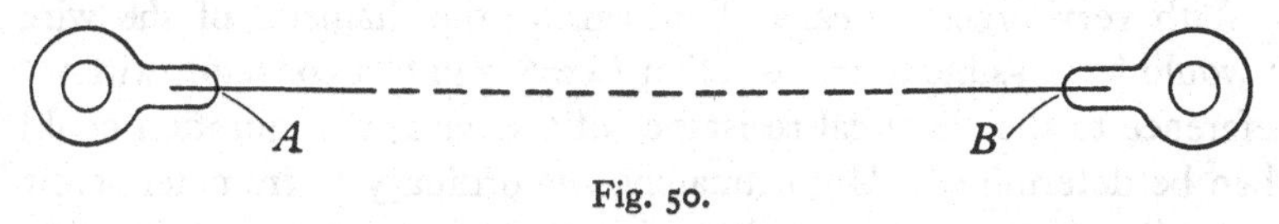

Fig. 50.

Take a length of about 50 cm. of insulated copper wire, of 24 S.W.G.; any thickness near that will serve. Bare its ends and solder to each of them a terminal tag, similar to those used in wireless work, but larger; there must be a hole in each tag, large enough to take a thick wire nail. Ample solder should be used, and it must cover the points A and B; the wire should point to the centre of the hole, since a sharp bend at A or B may cause the wire to snap.

Drive a nail into some firmly fixed object, and slip one tag over it; put a large thick wire nail through the hole in the other tag, to act as a handle. Put a metre scale under the wire and pull on the nail through the tag with a force which is gently increased until it stretches the wire through 2 or 3 mm.; this will smooth out any kinks and give a small permanent set. Measure the length (l cm.) of wire between A and B.

In one of the arms of a Wheatstone's bridge put a roughly equal length of wire of the same gauge, to act as a fixed resistance (call it R_0), and find the ratio of the resistance (R) of the wire between tags to this fixed resistance.

Repeat the careful stretching of the wire between tags, to give an increase in the permanent set, about 1 cm. long. Measure l and R/R_0 for this stretched wire.

Repeat, for a series of extensions of about 1 cm. each, until you have about ten readings or the wire breaks.

Plot a graph connecting l^2 and R/R_0. It will probably be a straight line through the origin. If so, you will have shown that the ratio of R to l^2 is constant for all the extensions you have made.

Denote by r the radius of the wire corresponding to the length l; then if the volume is unchanged by the extension, $l \times \pi r^2$ is constant. If the specific resistance is unchanged by the extension $\frac{R \times \pi r^2}{l}$ is constant. Hence if both volume and specific resistance are constant, R/l^2 is constant. You have probably found this to be so.

But that does not necessarily prove that both volume and specific resistance are constant; it is just possible, though unlikely, that there are changes in one that exactly balance changes in the other.

With very precise means of measuring the diameter of the wire it would be possible to test whether $l \times \pi r^2$ is in fact constant, without reference to the electrical resistance of the wire; this question could then be determined. Unfortunately, the ordinary micrometer screw gauge is not accurate enough to do so; in fact, on the assumption that neither volume nor specific resistance are changed by extension, this electrical method provides us with an exceptionally sensitive method of measuring the ratio of the radius of the stretched wire to the radius of the unstretched wires; we have seen that $lr^2 = l_0 r_0^2$ and $\frac{Rr^2}{l} = \frac{R_0 r_0^2}{l}$; hence by multiplying these we get $Rr^4 = R_0 r_0^4$ and $\frac{r}{r_0} = \left(\frac{R_0}{R}\right)^{\frac{1}{4}}$. R and R_0 can, of course, be measured to a high degree of accuracy.

But if you can stretch the wire so much as to increase its length by about 20 % without breaking it, the change in diameter will be large enough for a screw gauge to measure it roughly, and you can then check roughly the last formula.

Exp. 45. E.M.F. *in galvanometer circuit of a Wheatstone's bridge, I*

In a Wheatstone's bridge net, used for the determination of the ratio X/R by adjusting Y/Z so that no current is indicated by the galvanometer G, the result is not affected by any resistances R_1 and R_2 which there may be in the battery and galvanometer circuits, as shown in Fig. 51*a*. But an E.M.F. in the galvanometer circuit will affect the result; the accuracy attainable by this method is so high that even a thermo-electric effect in that circuit may be appreciable, so that we ought to be able to estimate the extent of its influence. In this and the two following experiments this is investigated step by step, to discover how the position of the balancing point P on the bridge wire is displaced when an E.M.F. of E volts is introduced into the galvanometer circuit, and how for a given value of E this displacement depends on R_1 and R_2 and on $X+R$ and on $Y+Z$.

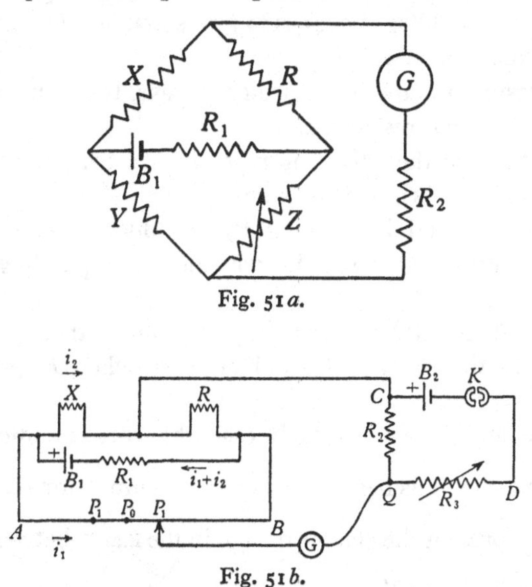

Fig. 51 *a*.

Fig. 51 *b*.

The set-up in Fig. 51 *b* is suitable for investigating this effect. B_1, B_2 are single accumulators with positive poles as shown, AB the bridge wire of length L_0 cm. and resistance R_0 ohms, G the galvanometer and CD a potentiometer.

This potentiometer is a device for producing between the points C and Q an E.M.F. in either direction which is a comparatively small measurable fraction of B_2. It may consist, as shown in Fig. 51 *b*, of a fixed resistance R_2 ohms between C and Q, and a resistance box giving R_3 ohms between Q and D, and a plug key K; or an ordinary potentiometer wire CQD can (less well) be used. For this experiment R_2 may well be a 1-ohm coil or a fixed length of about 25 cm. of resistance wire of about S.W.G. 28. Since, when there is no current through G, the P.D. between C and Q is $\frac{R_2}{R_2+R_3} B_2$ and B_2 is about 2 volts, a range of R_3 from 4 to 50 ohms will give a range of E.M.F.

(denote it by E) of about 0·4 to 0·04 volt; if K is open, E is zero, and if B_2 is reversed, the sign of E is changed.

R_1 may be a coil of about 5 ohms; if this equals the resistance R_0 of the bridge wire and also the sum of X and R, the P.D. between A and B will be about $\frac{2}{3}$ volt. X and R may be rough 2-ohm coils, or each may be about 25 cm. of S.W.G. 32 resistance wire. In this experiment we do not need to know the values of R_1, X and R, but they should be such (as above) as to give suitable values of P_0P_1 with the given potentiometer.

First measure, in any of the ordinary ways, the value of R_2 unless it is a coil of known resistance.

Find and record the balancing point P_0 (i.e. the point which gives no deflection in G) when K is open.

Then close K and by varying R_3 introduce a series of E.M.F.'s ($\pm E$ volts) into CQ and observe the corresponding balancing points P_1.

Deduce and record the values of P_0P_1, counting them positive when P_1 is, say, to the right of P_0 and negative when P_1 is to the left of P_0.

Calculate and record in each case the corresponding value of $\frac{R_2}{R_2+R_3}$, counting it as negative when the accumulator B_2 is reversed, in order to represent the change of sign in the E.M.F. between C and Q.

Plot a graph connecting P_0P_1 with $\frac{R_2}{R_2+R_3}$.

If, as is probable, this is a straight line passing through the origin, you will have found that the displacement of the balancing point is proportional to the magnitude of the introduced E.M.F., and reverses in sign with it, if all other quantities remain the same.

The appended note shows that this should theoretically be the case, and to this extent you will have checked the relation found therein:

$$P_0P_1 = L\frac{R_1}{B_1}\left(\frac{1}{R_0}+\frac{1}{R_1}+\frac{1}{X+R}\right)\frac{R_2}{R_2+R_3}B_2.$$

If your aim is merely to check this relation, it involves less calculation to plot the observed R_3 against the reciprocal of the observed P_0P_1, which can be taken directly out of a table of reciprocals. It will be seen from the theoretical equation that this should be a linear graph, not passing through the origin.

Note on the Theory

Denote by E_0 volts the fall in potential from A to B, then the P.D. between any two points P_0 and P_1 on the bridge wire (if it is uniform) is $\frac{P_0P_1}{L}E_0$ volts.

Hence if there is no current through G when it is connected to P_0 and K is open, and none when G is connected to P_1 and K is closed, the closure of K must introduce a P.D. of $\frac{P_0P_1}{L}E_0$ volts into the galvanometer circuit. But the closure of K introduces a P.D. of $\frac{R_2}{R_2+R_3}B_2$ volts into CQ, so

$$\frac{P_0P_1}{L}E_0=\frac{R_2}{R_2+R_3}B_2,$$

or

$$P_0P_1=L\frac{R_2}{R_2+R_3}\frac{B_2}{E_0}. \qquad (1)$$

Again, if i_1, i_2 are the currents shown in Fig. 51*b*,

$$E_0=R_0i_1=(X+R)i_2=B_1-R_1(i_1+i_2).$$

Eliminating i_2 from these equations,

$$(X+R+R_1)i_2=B_1-R_1i_1$$

and

$$R_0i_1=\frac{(X+R)(B_1-R_1i_1)}{X+R+R_1},$$

or

$$R_0i_1\left(1+\frac{R_1}{X+R}\right)=B_1-R_1i_1,$$

or

$$B_1=i_1R_1R_0\left(\frac{1}{R_0}+\frac{1}{R_1}+\frac{1}{X+R}\right).$$

Hence

$$\frac{1}{E_0}=\frac{1}{R_0i_1}=\frac{R_1}{B_1}\left(\frac{1}{R_0}+\frac{1}{R_1}+\frac{1}{X+R}\right),$$

or from (1)

$$P_0P_1=L\frac{R_1}{B_1}\left(\frac{1}{R_0}+\frac{1}{R_1}+\frac{1}{X+R}\right)\frac{R_2}{R_2+R_3}B_2. \qquad (2)$$

Exp. 46. *E.M.F. in galvanometer circuit of a Wheatstone's bridge, II*

The next step in checking equation (2) of Exp. 45 is to find the influence of a resistance in the battery circuit on the displacement of the balancing point caused by a constant E.M.F. in the galvanometer circuit. We use the same set-up as in Exp. 45, except that the potentiometer should now produce a fixed E.M.F. in CQ, while R_1 must be variable. The simplest way to provide for this is to move the resistance box to R_1, and to use for R_2 about 10 cm. and for R_2 about 90 cm. of, say, S.W.G. 32 Eureka wire, or any lengths in about the same ratio, provided that they do not get too hot or run down the battery B_2.

If R_0 is about 4 ohms, X and R can each conveniently be a 2-ohm coil.

With any value of R_1 find the balancing point P_0 when K is open; changes in R_1 under these conditions should not affect the position of P_0. Find the positions of P_1, the balancing point when K is closed, on each side of P_0 by reversing B_2. Deduce the two values of the length P_0P_1, counting both as positive, and calculate their arithmetic mean.

Repeat for a series of observed values of R_1, and plot a graph connecting R_1 and P_0P_1. This will probably turn out to be a straight line, not passing through the origin.

The note appended to Exp. 45 shows that theoretically

$$P_0P_1 = L\frac{R_2B_2}{(R_2+R_3)B_1}\left\{1+\left(\frac{1}{R_0}+\frac{1}{X+R}\right)R_1\right\}.$$

This relation between P_0P_1 and R_1 represents a straight line since L, R_2, B_2, B_1, R_3, X and R are constant. If your graph was a straight line you will have confirmed this part of the theory. The equation shows that when $R_1=0$, $P_0P_1 = L\frac{B_2}{B_1}\frac{R_2}{R_2+R_3}$. We can measure B_1 and B_2 by a voltmeter, but they should be so nearly equal that we can take $\frac{B_2}{B_1}=1$; if the wire used for the potentiometer is uniform, $\frac{R_2}{R_2+R_3}$ is the ratio of the length of the wire in CQ to the total length in the potentiometer and it can be determined by measuring these lengths. We can measure P_0P_1 when we make $R_1=0$ and compare its observed and calculated values, and thus check this part of the theory by direct experiment. Or, we can do so without putting $R_1=0$, by producing the graph to cut the axis of P_0P_1 and reading off its intercept; this should be equal to the value of $L\frac{B_2}{B_1}\frac{R_2}{R_2+R_3}$ calculated as above.

Again, the equation shows that $P_0P_1 = 0$ when

$$R_1 = -\frac{1}{\frac{1}{R_0} + \frac{1}{X+R}}.$$

Since this is a negative resistance we cannot experimentally produce the conditions which would make P_0P_1 zero (except by reducing the E.M.F. in CQ to zero) so we cannot directly check this part of the theory. But we can do so indirectly, by comparing the intercept of our graph on the axis of R_1 with the above theoretical value, if we know R_0. Alternatively, if we do not know R_0, if we assume that the theory is correct we can deduce from the intercept of the graph the value of R_0, the resistance of the bridge wire. It is worth while to perform this last calculation, and put the result on record.

Exp. 47. *E.M.F. in galvanometer circuit of a Wheatstone's bridge, III*

The last step in checking (2) is to find out how the value of $X+R$ affects the displacement P_0P_1 of the balancing point for a given E.M.F. in the galvanometer circuit. We use the same set-up as in Exp. 45, except that R_1 is now kept constant, and may well be a coil of about 4 ohms, R a fixed resistance which may well be a piece of bare S.W.G. 32 Eureka wire from 10 to 25 cm. long giving about 1 or 2 ohms; X is a variable resistance, of the same wire which can be varied in length between the binding screws from about 75 cm. to 10 cm. We do not need to know the values in ohms of X and R_1 provided that the resistance wire is uniform, so that their resistances are directly proportional to their lengths.

Start with about 10 cm. for X and measure the length (x_1 cm.) of free wire between the binding screws; measure the free length (x_2 cm.) of R; record the value of $x_1 + x_2$. Find the balancing point P_0 when K is open, and P_1 (on either or both sides of P_0) when K is closed. Deduce and record the value of P_0P_1.

Alter X, measuring the new value of x_1; the positions of both P_0 and P_1 will be changed and must be determined. Record the new values of $x_1 + x_2$ and P_0P_1.

Obtain similarly a series of corresponding values of $x_1 + x_2$ and P_0P_1; plot a smooth graph connecting P_0P_1 and $\frac{1}{x_1 + x_2}$. It will probably be a straight line

The note appended to Exp. 45 shows that theoretically

$$P_0P_1 = L\frac{R_1}{B_1}\left(\frac{1}{R_0}+\frac{1}{R_1}+\frac{1}{X+R}\right)\frac{R_2B_2}{R_2+R_3},$$

where $\frac{1}{X+R}$ is proportional to $\frac{1}{x_1+x_2}$. This represents a straight line, not passing through the origin; this theoretical result can be checked by the graph.

Combining the results of Exps. 45, 46 and 47, the general form of the equation (2) can be wholly checked.

Exp. 48. *Kirchhoff's Laws*

Kirchhoff's Laws assert that, in a network like Fig. 52, in the circuit ABC,

$$E_1 = R_1i_1 - R_3i_3, \tag{1}$$

and that, considering the points A or C,

$$i_1 + i_2 + i_3 = 0, \tag{2}$$

the directions of the currents being as shown in the figure.

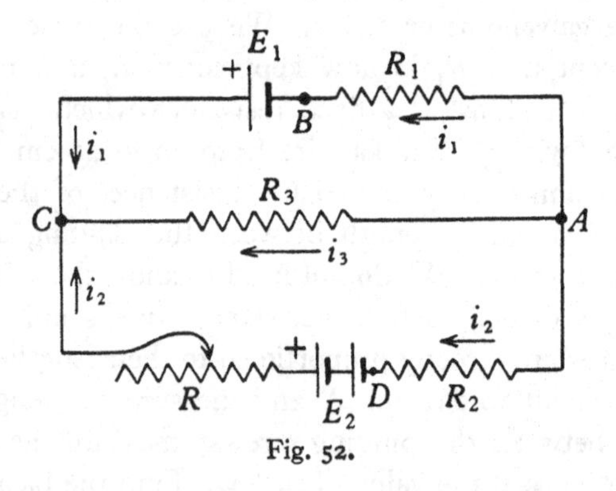

Fig. 52.

Dealing first with (1), if we accept Ohm's Law in the usual form, that the P.D. (V) across a conductor is proportional to the current (i) through it provided that the temperature is constant, and if we define the resistance (R) of the conductor as the value of this ratio, we get for each conductor $V/i = R$, or $Ri = V$.

It is important to specify definitely the signs to be attached to the P.D.'s between A and B, A and C, and A and D which we will denote numerically by V_1, V_2 and V_3 respectively. These should obviously be given the same signs as the respective currents so we must treat all three as positive when A has a higher potential than B, C or D; with these sign-conventions (1) becomes

$$E_1 = V_1 - V_3. \quad (3)$$

The truth of (3) is almost self-evident and stands in little need of experimental proof, for it merely expresses in symbols the fact that the P.D. between C and B is both E_1 and $V_1 - V_3$. But it is worth while to measure with a voltmeter the P.D. (V_1) between A and B, and the P.D. (V_3) between A and C, giving the observed signs to V_1 and V_3, and compare the results with the P.D. E_1 between B and C, if only because V_1 will change its sign during the experiment, and (3) should nevertheless hold good; this will be checked later.

Equation (2) is not so self-evident, and needs experimental verification.

The network can with advantage be made up with a single accumulator for E_1, two accumulators in series for E_2, and resistances made of lengths of S.W.G. No. 26 insulated resistance wire such as Eureka, which can carry 2 amps. without getting unduly hot if coiled in a loose spiral. R_1 and R_3 should each be about 60 cm. of this wire, giving roughly 2 ohms each, and R_2 about 15 cm. of it, giving roughly 0·5 ohm. R may be a rheostat capable of carrying 2 amps. and giving up to 3 ohms, or a 90 cm. length of the bared No. 26 resistance wire, to any point of which a lead from C can be attached by means of a connector or binding screw. Leads should be fixed to A, B, C and D so that either terminal of a voltmeter, reading up to 3·0 volts, can be connected to them as required.

The first step is to determine the values of R_1, R_2 and R_3. Since we may assume that the wire is of uniform resistance, if we measure the lengths we shall get numbers proportional to the resistances and that would suffice to test the theory, since the resistance in ohms of 1 cm. of the wire would cancel out of the equations. But it is more convenient to deal in ohms, and to take from the tables of constants the resistance per metre of the wire in use (which is 2·9 ohms for S.W.G. 26 Eureka) and hence find the approximate values of R_1, R_2 and R_3 in ohms. From them and the observed V_1, V_2 and V_3 in volts we can then deduce i_1, i_2 and i_3 in amperes.

We have agreed that V_1, V_2 and V_3 are positive if the potential of A is higher than that of B, D and C respectively; let us assume that i_1, i_2 and i_3 have the directions shown in Fig. 52; then V_1, V_2 and V_3 will all be positive.

Setting the rheostat to give its maximum resistance, measure V_1, V_2 and V_3, and deduce i_1, i_2 and i_3, giving to each its observed or deduced sign. Make a table of the results, and add columns to show $V_1 - V_3$ and $i_1 + i_2 + i_3$.

Reduce R step by step, until it is zero or until V_1, V_2 or V_3 exceeds the range of the voltmeter, whichever happens first, and fill up the table.

You will probably find that the last column is nearly zero in each case.

Measure with the voltmeter the E.M.F. E_1; you will probably find that this is nearly equal to the figures in the last column but one.

Thus you can test (2) and (3).

But if the single accumulator has just been fully charged, or if it is very nearly run down, and if your experiment lasts a long time, you may find that the values in the $V_1 - V_3$ column differ from one another. During the first part of the experiment this accumulator is discharging and during the second part it is being charged; towards the end of both processes its E.M.F. (E_1) changes fairly rapidly with the time. But it may show appreciably the same forward and backward E.M.F. at any stage of its charge or discharge, if it is switched over instantaneously from one state to the other.

Note on the Theory

If we apply Kirchhoff's Laws to the network of Fig. 52, in which $E_1 = 2$ volts, $E_2 = 4$ volts, we can deduce by elementary algebra the following expressions for the currents:

$$i_1 = \frac{2(R + R_2 - R_3)}{(R_1 + R_3)(R + R_2) + R_1 R_3},$$

$$i_2 = \frac{2(2R_1 + R_3)}{(R_1 + R_3)(R + R_2) + R_1 R_3},$$

$$i_3 = \frac{-2(R + 2R_1 + R_2)}{(R_1 + R_3)(R + R_2) + R_1 R_3}.$$

If, as in this experiment, $R_1 = 2$, $R_2 = 0{\cdot}5$ and $R_3 = 2$ ohms we get

$$i_1 = \frac{2R - 3}{4R + 6}, \quad i_2 = \frac{12}{4R + 6} \quad \text{and} \quad i_3 = \frac{-(2R + 9)}{4R + 6}.$$

Hence, for the following values of R we get, approximately,

R	i_1	i_2	i_3	V_1	V_2	V_3
3	0·167	0·667	−0·833	0·333	0·333	−1·667
2	0·071	0·856	−0·927	0·142	0·428	−1·856
1	−0·100	1·20	−1·10	−0·200	0·600	−2·20
0	−0·500	2·00	−1·50	−1·0	1·00	−3·00

It will be seen that a voltmeter reading 0–3 volts should cover the full range of observations.

Exp. 49. *Uniformity of a bridge wire*

Accuracy of the results when a wire bridge is used depends on the wire having a uniform resistance throughout its length; this is not always the case with a new instrument, and after long rough usage it may well not be the case. It can be tested by the following method, provided that the accuracy of the resistance box to be used can be trusted.

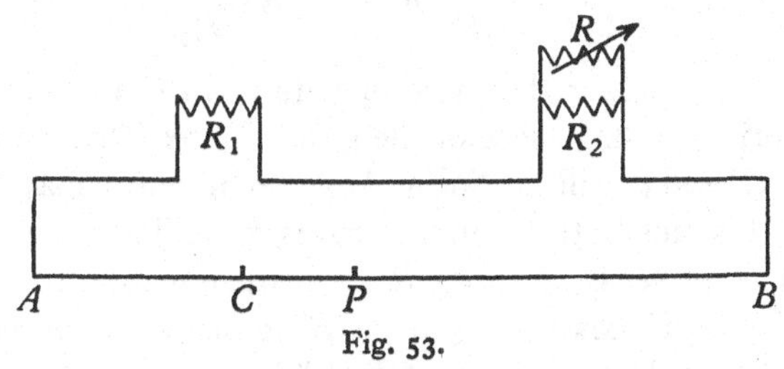

Fig. 53.

Cut off a length of insulated resistance wire, of resistance about 1 ohm; cut off another length about four times as long. Denote their resistances by R_1, R_2; put them in the gaps of a wire bridge, as in Fig. 53, and find the balancing point C.

In parallel with R_2 put a resistance box or separate resistances, and find the balancing point P for various values of R up to about 10 ohms, keeping R_1 and R_2 unchanged.

Plot a graph of $1/CP$ against R.

Theory shows that, if the resistance of the bridge wire is uniform between C and some point P_0 between C and B, the graph will be linear for the range of CP between O and CP_0, however irregular the bridge wire outside of CP_0 may be; and further, that if the resistance of the bridge wire is uniform from C to B, the reciprocal of the intercept of the graph on the $1/CP$ axis will be equal to CB.

Use your graph to test the uniformity of your bridge wire between C and B.

Note on the Theory

Denote by R_3 the resistance of AC, and by R_4 the resistance of CB; R_3 and R_4 are not necessarily proportional to the lengths. Denote by aCP the resistance of CP; a is not necessarily a constant for different positions of P.

Then
$$\frac{R_4}{R_3}=\frac{R_2}{R_1} \quad \text{or} \quad \frac{1}{R_2}=\frac{R_3}{R_1R_4}.$$

And
$$\frac{R_4-aCP}{R_3+aCP}=\frac{1}{\frac{1}{R}+\frac{1}{R_2}}\Bigg/R_1=\frac{1}{\frac{1}{R}+\frac{R_3}{R_1R_4}}\Bigg/R_1=\frac{RR_4}{R_1R_4+RR_3}.$$

Hence
$$\frac{R_4/aCP-1}{R_3/aCP+1}=\frac{RR_4}{R_1R_4+RR_3}.$$

Therefore
$$\frac{1}{CP}=\frac{a}{R_1R_4^2}(R_3+R_4)R+\frac{a}{R_4}.$$

If a is constant over the piece of wire from C to some point on CB, the corresponding piece of the graph represented by this equation in $1/CP$ and a will be linear, however much it may be curved outside of this range. If the graph proves to be linear over the range from C to B, we see by putting $R=0$ in the equation that the intercept on the $1/CP$ axis is a/R_4 or $1/CB$, so that the reciprocal of this intercept should theoretically be CB. This last is, of course, obvious, but in practice the intercept of the graph can only be obtained accurately by extrapolation.

Exp. 50. *Effect of an oblique magnet on a magnetometer*

In the following experiment the relation between the deflection of a magnetometer and the bearing, at the centre of the magnetometer, of a short bar magnet pointing towards that centre and at a constant distance from it, is investigated, and the theory checked.

If O is the centre of a magnetometer, with a needle of negligible length $2L$, and S_1N_1 a bar magnet of magnetic moment M and length $2l$, with its centre at a distance d from O and its axis pointing towards O, and if the bar magnet and the needle of the magnetometer, when it comes to rest, are inclined at angles of ϕ and θ to the meridian, theory shows that to a first approximation $M=\dfrac{(d^2-l^2)^2}{2d}H\dfrac{\sin\theta}{\sin(\phi-\theta)}$.

Take a deflection magnetometer, with an arm, and a short but strong bar magnet. Set the arm at right angles to the magnetic meridian and put the bar magnet on the arm at such a distance from O that $\theta = 30°$ approximately; record the reading of S_1, so that the magnet can be replaced in the same position when necessary.

Remove the bar magnet, and turn the magnetometer casing with its arm, so that ϕ takes the value of 15° approximately, using the pointer readings of the magnetometer needle before and after the movement to measure ϕ.

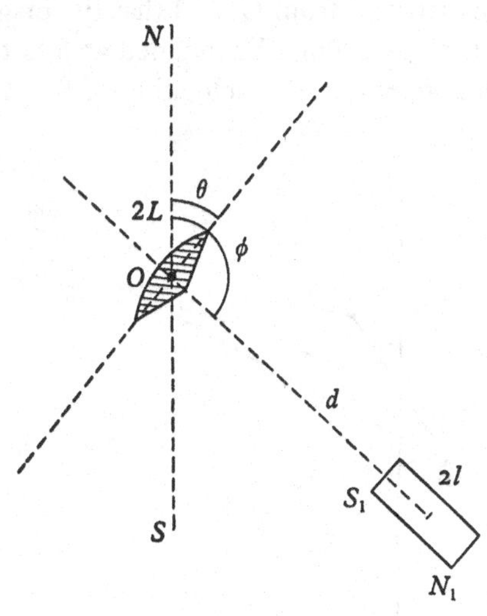

Fig. 54.

Replace the bar magnet at its former position, and measure θ by using the change in the pointer reading caused by the bar magnet.

Repeat with values of ϕ of 30°, 45°, ..., 150° approximately.

Plot a graph connecting $\sin\theta$ and $\sin(\phi-\theta)$. If this graph is a straight line through the origin, except for divergences which can be legitimately ascribed to errors of observation, your experiment confirms one part of the theory, and shows that in your case the lengths L and l were small enough in comparison with d to justify the approximations made.

The expression found for M in the appended note on the theory shows that any errors made in estimating the effective length (l) of the bar magnet will not affect the straightness of your graph, though they will affect its slope; but if $(L/d)^2$ is not negligibly small

$$\frac{\sin\theta}{\sin(\phi-\theta)}\left\{1-\frac{15L^2\cos^2(\phi-\theta)}{2d^2}\right\}$$

is a constant, and not $\dfrac{\sin\theta}{\sin(\phi-\theta)}$, so that the graph connecting $\sin\theta$ and $\sin(\phi-\theta)$ will be curved and not straight. The consequent curvature will, however, be small.

Note on the Theory

Suppose that we have a needle of length $2L$, with poles of strength m', pivoted at O, and a bar magnet of length $2l$ with poles of strength m, with its centre distant d from O, and that the magnet's axis passes through O. Let the bar magnet be inclined at ϕ to the meridian and let the needle come to rest at an inclination of θ to the meridian.

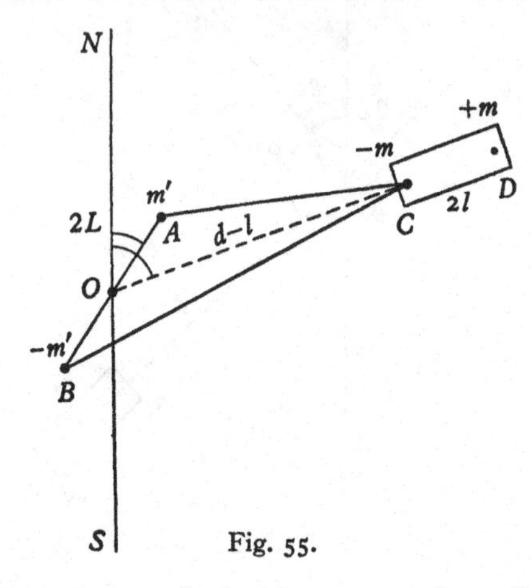

Fig. 55.

The pole at C produces a clockwise moment about O on the needle of

$$\frac{Lmm'}{CA^2}\sin CAO+\frac{Lmm'}{CB^2}\sin CBO$$

$$=Lmm'\left\{\frac{d-l}{CA^3}\sin(\phi-\theta)+\frac{d-l}{CB^3}\sin(\phi-\theta)\right\}$$

$$=Lmm'(d-l)\sin(\phi-\theta)\left[\{L^2+(d-l)^2-2L(d-l)\cos(\phi-\theta)\}^{-\frac{3}{2}}+\{L^2+(d-l)^2+2L(d-l)\cos(\phi-\theta)\}^{-\frac{3}{2}}\right].$$

Treating l and L as small compared with d this becomes

$$\frac{Lmm'(d-l)\sin(\phi-\theta)}{(d-l)^3}\left[\left\{1-\frac{2L}{d}\cos(\phi-\theta)\right\}^{-\frac{3}{2}}+\left\{1+\frac{2L}{d}\cos(\phi-\theta)\right\}^{-\frac{3}{2}}\right]$$

or $$\frac{Lmm'\sin(\phi-\theta)}{(d-l)^2}\left[1+\frac{15}{2}\frac{L^2}{d^2}\cos^2(\phi-\theta)+1+\frac{15}{2}\frac{L^2}{d^2}\cos^2(\phi-\theta)\right]$$

or $$\frac{2Lmm'\sin(\phi-\theta)}{(d-l)^2}\left\{1+\frac{15}{2}\frac{L^2}{d^2}\cos^2(\phi-\theta)\right\}.$$

The pole $+m$ at D produces a counter-clockwise moment as above, but with $-l$ changed to $+l$. So the bar magnet produces a clockwise moment of

$$2Lmm' \sin(\phi-\theta)\left\{1+\frac{15}{2}\frac{L^2}{d^2}\cos^2(\phi-\theta)\right\}\left\{\frac{1}{(d-l)^2}-\frac{1}{(d+l)^2}\right\},$$

or

$$\frac{8dlLmm' \sin(\phi-\theta)}{(d^2-l^2)^2}\left\{1+\frac{15}{2}\frac{L^2}{d^2}\cos^2(\phi-\theta)\right\}.$$

The counter-clockwise moment on the needle exerted by the Earth's field of magnetic force is $2Lm'H\sin\theta$. These must be equal, so

$$\frac{4dMLm' \sin(\phi-\theta)}{(d^2-l^2)^2}\left\{1+\frac{15}{2}\frac{L^2}{d^2}\cos^2(\phi-\theta)\right\}=2Lm'H\sin\theta,$$

where M is the magnetic moment of the bar magnet, or

$$M=\frac{(d^2-l^2)^2}{2d}H\frac{\sin\theta}{\sin(\phi-\theta)}\left\{1-\frac{15}{2}\frac{L^2}{d^2}\cos^2(\phi-\theta)\right\}.$$

If $\phi=90°$ this reduces to the ordinary formula for the A position of Gauss.

APPENDIX A

Parallax Methods with Lenses

According to the most elementary theory of lenses the real image of a point more than the focal length behind a convergent lens is another single point in front of the lens, and all rays from the former point which pass through the lens pass through the latter point. If that were true it would be easy to find the position of the latter point by means of a search pin, by moving the search pin about until a position is found where the pin seems to coincide with the image of an object pin at the former point, whatever be the position of the eye. There is then said to be no 'parallax' between the search pin and the image of the object pin; this 'parallax method' is in constant use in optical experiments, and in many cases it gives a reasonably accurate way of locating an image.

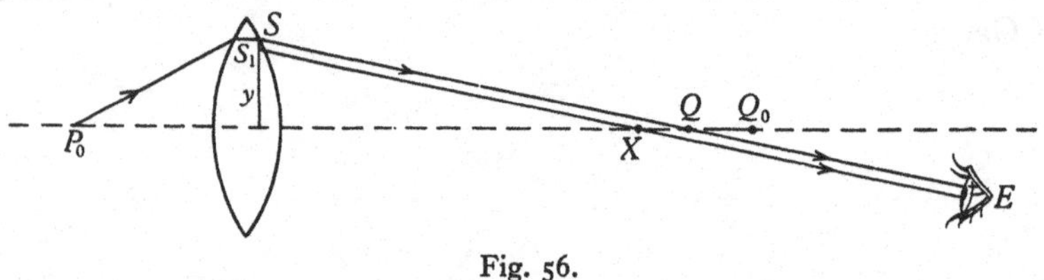

Fig. 56.

But in some cases it fails badly, and in nearly all cases it calls for the exercise of intelligent judgement to get a reasonably accurate result. The following discussion of the way in which the unavoidable spherical aberration of a lens affects this parallax method, and of the way in which its effects can be allowed for, may help the experimenter to exercise that intelligent judgement.

Suppose that the point of an object pin is put at P_0 on the optic axis of a convergent lens and that a single eye is put at E, at a considerable distance from the lens and at some little distance from its optic axis. Looking at the lens this eye will see an image of P_0 apparently situated at S on the face of the lens; denote the distance OS of S from the optic axis by y. If the eye starts in the optic axis and moves broadside away from that axis, in either direction, S will seem to move outwards from the centre of the lens in the opposite direction, so that the value of y depends on the position of the eye.

Suppose that the ray from S to E cuts the optic axis at Q. Theory and practice (as exemplified in Exp. 20) show that Q moves along the optic axis as the eye at E moves across that axis; as E approaches more nearly to the axis, Q approaches more nearly to the point Q_0, the focus conjugate to P_0 for a thin *axial* pencil proceeding from P_0. For any distance y of S from the axis, theory and practice (as in Exp. 20) show that $QQ_0 = Ay^2$, where A is very nearly constant. QQ_0 is often called the spherical aberration of the lens.

Now suppose that the point of a search pin is put at X in the figure. The eye at E will see X apparently projected on the face of the lens, at S_1 say, along the line EXS_1, which is parallel to EQS. The separation SS_1 can conveniently be termed the 'parallax' seen by the eye at E between the search pin and the image of P_0. We will regard this parallax as positive if S_1 is farther from O than S, so it will be negative in the figure.

Theory shows that the parallax SS_1 is approximately equal to $\frac{y(Ay^2 - XQ_0)}{v_0}$, where v_0 is the effective distance of Q_0 from the lens (or Q_0H_1, where H_1 is a principal point of the lens, see Exp. 30).

Let us apply this to the ordinary case of a convergent lens made of glass of $\mu = 1{\cdot}5$, with spherical faces of equal curvature, and a focal length of f; if $P_0O = u_0$ and $OQ_0 = v_0$, theory shows that $A = \frac{v_0^2}{6f}\left(\frac{10}{f^2} - \frac{13}{u_0 v_0}\right)$.

Consider the particular case when $f = 10$ cm., $u_0 = 15$ cm. and therefore by the ordinary lens formula $v_0 = 30$ cm. Suppose that the lens has a comparatively large diameter, say 5·0 cm. Then the above formula gives $A = 1{\cdot}07$, and the ordinates of the curves in Fig. 57 represent the parallax, SS_1 in millimetres, calculated from its foregoing value, for values of y as abscissae if XQ_0 has the values in centimetres noted against each curve.

It will be seen that if we put the search pin at Q_0, the point for which we are searching, the parallax, as shown in the curve labelled O, increases continuously from zero, slowly at first and more rapidly later; so the search pin always appears to be outside the image on the lens. The same is true if the search pin has been put 2 or 3 cm. farther from the lens than Q_0 so that $XQ_0 = -2$ or -3; but here the parallax curves cut and do not touch the axis of y, whereas the curve for $XQ_0 = O$ is a tangent to it. If the search pin is put 2 or 3 cm. nearer to the lens than Q_0, the parallax starts at zero, increases numerically to a negative maximum, becomes zero and then becomes positive and increases continuously. So the appearance on the lens in these latter cases is that the search pin appears at first to lag behind the image on the lens as the eye moves outwards from the optic axis, then catches it up, passes it and afterwards gains rapidly on it. This can easily be seen to be the case if the apparatus is set up as in Exp. 20, and it is illustrated in Fig. 21*a*.

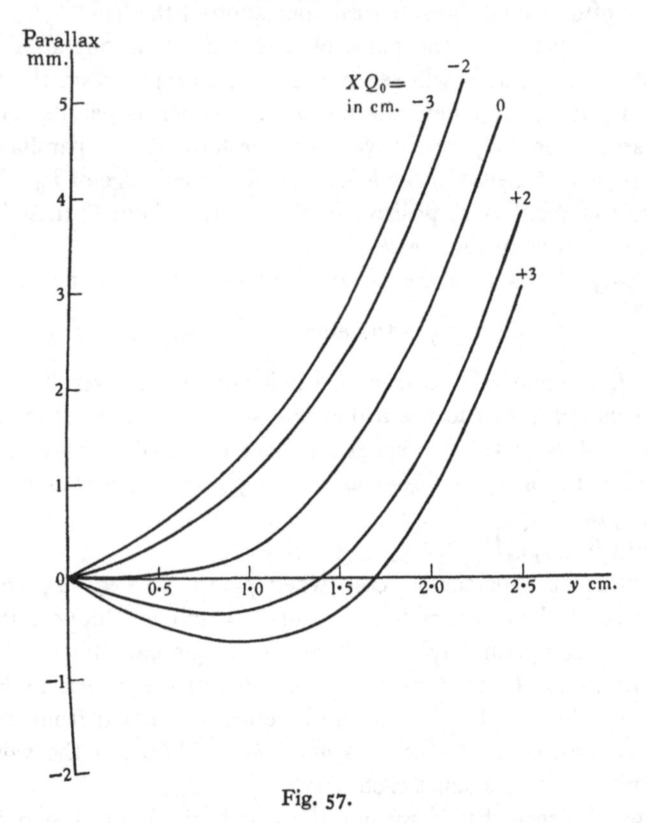

Fig. 57.

A succession of pairs of relative positions of search pin and image is there shown, as the eye moves outward from the optic axis in either direction.

We can see from the curves in Fig. 57 that there are two available methods of locating Q_0. The first is an indirect method, making use of the points where the parallax curves cut the y axis, that is, the values of y for which the parallax vanishes; it was explained and used in Exp. 20. The second is a direct method, to feel about with a search pin until the image and search pin behave, on the lens, as they should when the search pin is at Q_0; that is, the parallax must change as it does in the parallax curve labelled *O* in the figure. It is not easy in practice to distinguish between the behaviours indicated by these curves; the total parallax observable is often quite small, and the curves 'below the line' are then close to the axis of *y* for a considerable distance across the lens, while the curve labelled *O* is also close to the axis for a considerable distance. The result is that the search pin is generally set too near the lens, a temptation to be resisted.

It will be seen that the error of setting, that is XQ_0, may be of substantial amount, 2 or 3 cm., in short focus lenses such as that we have been considering.

Either of these parallax methods can be used in the location of a principal focus by means of a plane mirror and a pin.

The subject is treated more fully in *Phil. Mag.*, Vol. 38, No. 283 for August 1947, and the *School Science Review*, Vol. XXIX, No. 108 for March 1948.

APPENDIX B

Thick Lenses

In this appendix we will deal only with conjugate foci formed by thin *axial* pencils of light.

In the foregoing pages the convention of signs usually called 'Real is Positive' has been used. One of the reasons for this choice is that the majority of students is accustomed to it, even though some of them may have adopted another convention at a later stage of their course; another reason is that practical work in optics deals for the most part with real images, and most of the quantities to be measured are positive in that convention, so that it economises thought to adopt it.

But the theoretical proofs in this appendix are sufficiently complicated to call for the use of the most convenient sign convention, which is that one sometimes called the 'New Cartesian', in which the positive direction is that in which light is moving, with all diagrams drawn so that the positive direction is from left to right. Although the form of the proofs depends intimately on the sign-convention adopted, it will be seen that the results obtained are so simple that they can very easily be modified to suit any other sign convention, so that they can be used in practical work, whatever convention is used there.

(i) *Single lens*

Take a lens of thickness $A_1A_2 = t$, with radii of curvature r and s and index of refraction μ. Suppose that rays emanate from P_0 and that q and P_1 are the conjugate foci after refraction at the two faces.

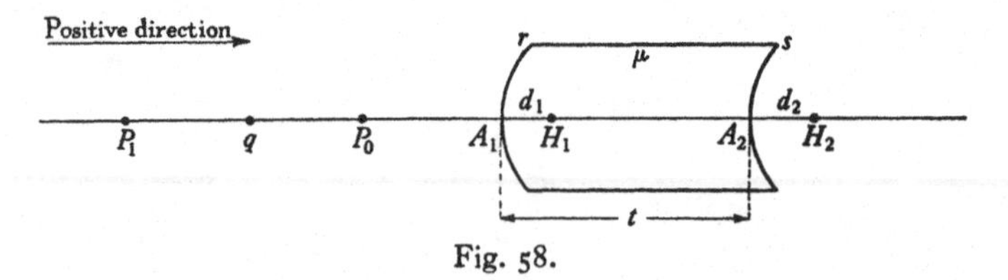

Fig. 58.

Let u_1 and w_1 denote the distances of P_0 and q respectively to the right of A_1; then in Fig. 58 $A_1P_0 = -u_1$ and $A_1q = -w_1$. By the elementary theory of refraction of light at a spherical surface

$$\frac{\mu}{w_1} - \frac{1}{u_1} = \frac{\mu - 1}{r}.$$

Let w_2 and u_2 denote the distances of q and P_1 respectively to the right of A_2; then in Fig. 58

$$A_2q = -w_2 \quad \text{and} \quad A_2P_1 = -u_2.$$

Then
$$\frac{\frac{1}{\mu}}{u_2} - \frac{1}{w_2} = \frac{\frac{1}{\mu}-1}{s} \quad \text{or} \quad \frac{\mu}{w_2} - \frac{1}{u_2} = \frac{\mu-1}{s}.$$

But from Fig. 58,

$$t = A_1A_2 = A_2q - A_1q = -w_2 + w_1.$$

Substituting for w_2 in the second equation we get

$$\frac{\mu}{w_1 - t} = \frac{1}{u_2} + \frac{\mu-1}{s}.$$

Hence the two equations can be written

$$\frac{w_1}{\mu} = \frac{1}{\frac{1}{u_1} + \frac{\mu-1}{r}} \quad \text{and} \quad \frac{w_1}{\mu} - \frac{t}{\mu} = \frac{1}{\frac{1}{u_2} + \frac{\mu-1}{s}}.$$

Eliminating $\frac{w_1}{\mu}$,
$$\frac{t}{\mu} = \frac{1}{\frac{1}{u_1} + \frac{\mu-1}{r}} - \frac{1}{\frac{1}{u_2} + \frac{\mu-1}{s}}$$

or
$$\frac{t}{\mu}\left(\frac{1}{u_1} + \frac{\mu-1}{r}\right)\left(\frac{1}{u_2} + \frac{\mu-1}{s}\right) = \frac{1}{u_2} - \frac{1}{u_1} - (\mu-1)\left(\frac{1}{r} - \frac{1}{s}\right).$$

This equation can be simplified by changing the origins from which foci, P_0, etc., are measured. Take two arbitrary points, H_1 and H_2, respectively d_1 and d_2 to the right of A_1 and A_2, as shown, and denote the distance of P_0 to the right of H_1 by u, and the distance of P_1 to the right of H_2 by v. Then in the figure, $H_1P_0 = -u$ and $H_2P_1 = -v$, and

$$u = -H_1P_0 = -(A_1H_1 + A_1P_0) = -d_1 + u_1.$$

Hence $u_1 = u + d_1$, and similarly $u_2 = v + d_2$.

Substituting these values and multiplying both sides by

$$\left(1 + \frac{d_1}{u}\right)\left(1 + \frac{d_2}{v}\right),$$

we get

$$\frac{t}{\mu}\left\{\frac{1}{u} + \frac{\mu-1}{r}\left(1 + \frac{d_1}{u}\right)\right\}\left\{\frac{1}{v} + \frac{\mu-1}{s}\left(1 + \frac{d_2}{v}\right)\right\}$$
$$= \frac{1}{v}\left(1 + \frac{d_1}{u}\right) - \frac{1}{u}\left(1 + \frac{d_2}{v}\right) - (\mu-1)\left(\frac{1}{r} - \frac{1}{s}\right)\left(1 + \frac{d_1}{u} + \frac{d_2}{v} + \frac{d_1d_2}{uv}\right)$$

or $$\frac{1}{v}\left\{1-(\mu-1)\left(\frac{1}{r}-\frac{1}{s}\right)d_2-\frac{t(\mu-1)}{\mu v}-\frac{t(\mu-1)^2 d_2}{\mu rs}\right\}$$

$$-\frac{1}{u}\left\{1+(\mu-1)\left(\frac{1}{r}-\frac{1}{s}\right)d_1+\frac{t(\mu-1)}{\mu s}+\frac{t(\mu-1)^2 d_1}{\mu rs}\right\}$$

$$+\frac{1}{uv}\left\{d_1-d_2-(\mu-1)\left(\frac{1}{r}-\frac{1}{s}\right)d_1 d_2-\frac{t}{\mu}-\frac{t(\mu-1)d_1}{\mu r}\right.$$

$$\left.-\frac{t(\mu-1)d_2}{\mu s}-\frac{t(\mu-1)^2 d_1 d_2}{\mu rs}\right\}$$

$$=(\mu-1)\left(\frac{1}{r}-\frac{1}{s}\right)+\frac{(\mu-1)^2 t}{\mu rs}.$$

Denote $$(\mu-1)\left(\frac{1}{r}-\frac{1}{s}\right)+\frac{(\mu-1)^2 t}{\mu rs} \text{ by } \frac{1}{f}. \tag{1}$$

Then the equation can be written

$$\frac{1}{v}\left\{1-\frac{(\mu-1)t}{\mu r}-\frac{d_2}{f}\right\}-\frac{1}{u}\left\{1+\frac{(\mu-1)t}{\mu s}+\frac{d_1}{f}\right\}$$

$$+\frac{1}{uv}\left\{d_1-d_2-\frac{d_1 d_2}{f}-\frac{t}{\mu}-\frac{t(\mu-1)d_1}{\mu r}-\frac{t(\mu-1)d_2}{\mu s}\right\}=\frac{1}{f}.$$

For brevity, denote this equation by $\frac{A}{v}-\frac{B}{u}+\frac{K}{uv}=\frac{1}{f}$. Then, multiplying out $A\times B$,

$$A\times B-1=\frac{d_1}{f}-\frac{d_2}{f}-\frac{d_1 d_2}{f^2}-\frac{(\mu-1)td_1}{\mu rf}-\frac{(\mu-1)td_2}{\mu sf}$$

$$-\frac{t}{\mu}\left\{(\mu-1)\left(\frac{1}{r}-\frac{1}{s}\right)+\frac{(\mu-1)^2 t}{\mu rs}\right\}$$

$$=\frac{d_1}{f}-\frac{d_2}{f}-\frac{d_1 d_2}{f^2}-\frac{t}{\mu f}-\frac{t(\mu-1)d_1}{\mu rf}-\frac{t(\mu-1)d_2}{\mu sf}, \text{ by (1).}$$

Hence it becomes obvious that $K=A\times B-1$, and the equation becomes

$$\frac{1}{v}\left\{1-\frac{(\mu-1)t}{\mu r}-\frac{d_2}{f}\right\}-\frac{1}{u}\left\{1+\frac{(\mu-1)t}{\mu s}+\frac{d_1}{f}\right\}$$

$$+\frac{f}{uv}\left\{\left(1-\frac{(\mu-1)t}{\mu r}-\frac{d_2}{f}\right)\left(1+\frac{(\mu-1)t}{\mu s}+\frac{d_1}{f}\right)-1\right\}=\frac{1}{f}. \tag{2}$$

There are two obvious choices of d_1 and d_2 for simplifying (2).

(A) We can make the coefficients of $\frac{1}{v}$ and $-\frac{1}{u}$ each equal to unity, by taking

$$d_1 = -\frac{(\mu-1)ft}{\mu s} \quad \text{and} \quad d_2 = -\frac{(\mu-1)ft}{\mu r}. \tag{3}$$

In that case the coefficient of $\frac{1}{uv}$ vanishes and (2) becomes

$$\frac{1}{v}-\frac{1}{u}=\frac{1}{f}, \tag{4}$$

where f is given by (1). This is Halley's or the 'ordinary' lens formula; f is then the focal length of the lens as ordinarily defined.

(B) We can make the coefficients of $\frac{1}{v}$ and $-\frac{1}{u}$ vanish, by taking

$$d_1 = -\left(f+\frac{(\mu-1)ft}{\mu s}\right) \quad \text{and} \quad d_2 = f-\frac{(\mu-1)ft}{\mu r}.$$

In that case the coefficient of f/uv is -1, and (2) becomes

$$uv = -f^2,$$

where f is given by (1). This is Newton's lens formula.

Both these formulae hold good for lenses of any thickness, since we have set no limit to the magnitude of t.

If we adopt Halley's formula the points H_1 and H_2 are called the Principal Points of the lens; we see by Fig. 58 that the distance between them is

$$\begin{aligned} H_1H_2 = A_1A_2 - d_1 + d_2 &= t+\frac{(\mu-1)ft}{\mu s}-\frac{(\mu-1)ft}{\mu r} \quad \text{by (3)} \\ &= t-\frac{ft}{\mu}(\mu-1)\left(\frac{1}{r}-\frac{1}{s}\right) \\ &= t-\frac{ft}{\mu}\left\{\frac{1}{f}-\frac{(\mu-1)^2 t}{\mu rs}\right\} \quad \text{by (1)} \\ &= \frac{(\mu-1)t}{\mu}+\frac{(\mu-1)^2 ft^2}{\mu^2 rs}. \end{aligned} \tag{5}$$

If the lens is so thin that the second term on the right-hand side is negligibly small, we can take $H_1H_2 = (\mu-1)t/\mu$. If, in addition, μ is nearly equal to 3/2, *the distance between the 'Principal Points' H_1 and H_2 is nearly one-third of the thickness of the lens.* If, in addition, the lens is biconvex or biconcave, so that $r=-s$, symmetry shows that H_1 and H_2 are each within the lens, at one-third of its thickness from the respective faces.

If, on the other hand, we adopt Newton's formula, u and v must be measured from the Principal Foci; in that case the symbols u and v are usually replaced by p and q to prevent confusion.

(ii) *Combination of two lenses, out of contact*

Suppose that we have two thick lenses arranged coaxially, with characteristics as shown in Fig. 59; denote A_2A_1' by a. Let H_1, H_2 and H_1', H_2' be their principal points, chosen to give Halley's lens formula for each of them separately, then by (3)

$$A_1H_1 = d_1 = -\frac{(\mu-1)ft}{\mu s}, \qquad A_2H_2 = d_2 = -\frac{(\mu-1)ft}{\mu r}, \tag{6}$$

$$A_1'H_1' = d_1' = -\frac{(\mu'-1)f't'}{\mu' s'}, \qquad A_2'H_2' = d_2' = -\frac{(\mu'-1)f't'}{\mu' r'}. \tag{7}$$

Let P_0 (u to the right of H_1) and q (v to the right of H_2) be conjugate foci for the first lens; let q (u' to the right of H_1') and P_1 (v' to the right of H_2') be conjugate foci for the second lens.

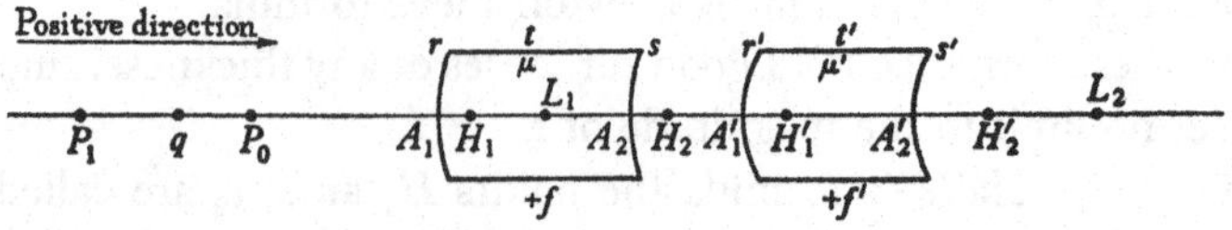

Fig. 59.

Then by (4)

$$\frac{1}{v} - \frac{1}{u} = \frac{1}{f} = (\mu-1)\left(\frac{1}{r} - \frac{1}{s}\right) + \frac{(\mu-1)^2 t}{\mu rs}, \tag{8}$$

and

$$\frac{1}{v'} - \frac{1}{u'} = \frac{1}{f'} = (\mu'-1)\left(\frac{1}{r'} - \frac{1}{s'}\right) + \frac{(\mu'-1)^2 t'}{\mu' r' s'}. \tag{9}$$

Now from Fig. 59

$$\begin{aligned} u' = -H_1'q &= -(H_2q + H_1'H_2) = -H_2q - (A_2H_1' - A_2H_2) \\ &= -H_2q - (A_2A_1' + A_1'H_1' - A_2H_2) \\ &= v - a + \frac{(\mu'-1)f't'}{\mu' s'} - \frac{(\mu-1)ft}{\mu r}. \end{aligned}$$

Denote
$$-a + \frac{(\mu'-1)f't'}{\mu' s'} - \frac{(\mu-1)ft}{\mu r} \text{ by } C. \tag{10}$$

Then
$$C = -H_1'H_2 \quad \text{and} \quad u' = v + c. \tag{11}$$

Take two arbitrary points, L_1 and L_2, respectively, d_3 and d_4 to the right of H_1 and H'_2, so that in Fig. 59 $H_1L_1 = d_3$ and $H'_2L_2 = d_4$. Suppose that P_0 is U to the right of L_1 and P_2 is V to the right of L_2. Then

$$u = -H_1P_0 = -(L_1P_0 - L_1H_1) = U + d_3,$$

and
$$v' = -H'_2P_1 = -(L_2P_1 - L_2H'_2) = V + d_4.$$

Hence (8) becomes
$$\frac{1}{v} - \frac{1}{U+d_3} = \frac{1}{f},$$

and (9) becomes
$$\frac{1}{V+d_4} - \frac{1}{v+C} = \frac{1}{f'}.$$

Eliminating v from these equations

$$\frac{1}{\dfrac{1}{U+d_3} + \dfrac{1}{f}} = \frac{1}{\dfrac{1}{V+d_4} - \dfrac{1}{f'}} - C$$

or
$$\frac{1}{V+d_4} - \frac{1}{f'} - \frac{1}{U+d_3} - \frac{1}{f} = -C\left(\frac{1}{V+d_4} - \frac{1}{f'}\right)\left(\frac{1}{U+d_3} + \frac{1}{f}\right).$$

Multiply both sides by $(1 + d_3/U)(1 + d_4/V)$. Then

$$\frac{1}{V}\left(1 + \frac{d_3}{U}\right) - \frac{1}{U}\left(1 + \frac{d_4}{V}\right) - \left(\frac{1}{f} + \frac{1}{f'}\right)\left(1 + \frac{d_3}{U} + \frac{d_4}{V} + \frac{d_3 d_4}{UV}\right)$$
$$= -C\left(\frac{1}{V} - \frac{1}{f'} - \frac{d_4}{f'V}\right)\left(\frac{1}{U} + \frac{1}{f} + \frac{d_3}{fU}\right)$$

or

$$\frac{1}{V}\left\{1 - d_4\left(\frac{1}{f} + \frac{1}{f'}\right) + \frac{C}{f}\left(1 - \frac{d_4}{f'}\right)\right\} - \frac{1}{U}\left\{1 + d_3\left(\frac{1}{f} + \frac{1}{f'}\right) + \frac{C}{f'}\left(1 + \frac{d_3}{f}\right)\right\}$$
$$+ \frac{1}{UV}\left\{d_3 - d_4 - \left(\frac{1}{f} + \frac{1}{f'}\right) d_3 d_4 + C\left(1 + \frac{d_3}{f} - \frac{d_4}{f'} - \frac{d_3 d_4}{ff'}\right)\right\} = \frac{1}{f} + \frac{1}{f'} + \frac{C}{ff'}.$$

Denote
$$\frac{1}{f} + \frac{1}{f'} + \frac{C}{ff'} \text{ by } \frac{1}{F}. \tag{12}$$

Then the equation becomes, by using (12),

$$\frac{1}{V}\left(1 + \frac{C}{f} - \frac{d_4}{F}\right) - \frac{1}{U}\left(1 + \frac{C}{f'} + \frac{d_3}{F}\right)$$
$$+ \frac{1}{UV}\left(d_3 - d_4 - \frac{d_3 d_4}{F} + C + \frac{Cd_3}{f} - \frac{Cd_4}{f'}\right) = \frac{1}{F}.$$

For brevity, denote this equation by $\dfrac{A}{V} - \dfrac{B}{U} + \dfrac{K}{UV} = \dfrac{1}{F}$. Then $A \times B - 1$ is

$$\frac{d_3}{F} - \frac{d_4}{F} + \frac{C}{f'} + \frac{C}{f} + \frac{C^2}{ff'} + \frac{Cd_3}{fF} - \frac{Cd_4}{f'F} - \frac{d_3 d_4}{F^2}$$

or, using (12) $$\frac{d_3}{F}-\frac{d_4}{F}-\frac{d_3 d_4}{F^2}+\frac{C}{F}+\frac{Cd_3}{fF}-\frac{Cd_4}{f'F}.$$

Obviously, $K=(A\times B-1)F$.

Hence the equation can be written in the form

$$\frac{1}{V}\left(1+\frac{C}{f}-\frac{d_4}{F}\right)-\frac{1}{U}\left(1+\frac{C}{f'}+\frac{d_3}{F}\right)$$
$$+\frac{F}{UV}\left\{\left(1+\frac{C}{f}-\frac{d_4}{F}\right)\left(1+\frac{C}{f'}+\frac{d_3}{F}\right)-1\right\}=\frac{1}{F}.$$

As in § (i), this can be simplified by taking

(A) $$d_3=-\frac{CF}{f'} \quad\text{and}\quad d_4=\frac{CF}{f}, \tag{13}$$

in which case the equation reduces to

$$\frac{1}{V}-\frac{1}{U}=\frac{1}{F},$$

where F is given by (12) and (10). Hence Halley's lens-formula is strictly true for a combination of two coaxial lenses of any thickness at any distance apart.

(B) If we take

$$d_4=F\left(1+\frac{C}{f}\right) \quad\text{and}\quad d_3=-F\left(1+\frac{C}{f'}\right)$$

the equation reduces to $UV=-F^2$.

Hence Newton's lens formula is strictly true for the same combination; the focal length F is given by (12) and (10). These results are applicable to the combinations of lenses which form simple telescopes, simple microscopes and compound eyepieces.

But it will be seen from (12) and (11) that if $f+f'=-C=H_2H_1'$, F becomes infinitely large, and the Principal Points and Principal Foci move off to infinity; this particular case needs special treatment, as in Exp. 32.

To determine the distance between the principal points, we see by (12) that $F=\dfrac{ff'}{f+f'+C}$, so that by (13) $H_1L_1=-\dfrac{Cf}{f+f'+C}$. Similarly $H_2'L_2=\dfrac{Cf'}{f+f'+C}$. Hence from Fig. 59

$$L_1L_2=H_1H_2-H_1L_1+H_2H_1'+H_1'H_2'+H_2'L_2.$$

Applying (5) and (11)

$$L_1L_2=\frac{(\mu-1)\,t}{\mu}+\frac{(\mu-1)^2\,ft^2}{\mu^2rs}+\frac{Cf}{f+f'+C}-C+\frac{(\mu'-1)\,t'}{\mu'}+\frac{(\mu'-1)^2\,f't'^2}{\mu'^2r's'}+\frac{Cf'}{f+f'+C}$$

$$=\frac{(\mu-1)\,t}{\mu}+\frac{(\mu'-1)\,t'}{\mu'}+\frac{(\mu-1)^2\,ft^2}{\mu^2rs}+\frac{(\mu'-1)\,f't'^2}{\mu'^2r's'}-\frac{C^2}{f+f'+C}. \tag{14}$$

(iii) *Two moderately thick lenses in contact*

Consider the particular case when $a=0$ and t, t' and C are so small that we can ignore their squares in comparison with $f+f'+C$, r, s, etc. Then L_1L_2 becomes

$$\frac{(\mu-1)\,t}{\mu}+\frac{(\mu'-1)\,t'}{\mu'}. \tag{15}$$

If, further, the lenses are biconvex or biconcave, i.e. if $r=-s$ and $r'=-s'$, then

$$\frac{1}{f}=\frac{2(\mu-1)}{r},\quad \frac{1}{f'}=-\frac{2(\mu'-1)}{s'}\quad\text{and}\quad C=-\frac{t'}{2\mu'}-\frac{t}{2\mu};$$

if, further, $\mu=\mu'=\frac{3}{2}$ nearly, then

$$\frac{1}{F}=\frac{1}{f}+\frac{1}{f'}-\frac{(t+t')}{3ff'}.$$

Hence in the determination of the focal length f' of a divergent lens of this kind, by measuring f of a stronger convergent lens of the same kind and F of the combination of the two in contact, the value of f' is given by

$$\frac{1}{f'}\left\{1-\frac{(t+t')}{3f}\right\}=\frac{1}{F}-\frac{1}{f}\quad\text{or}\quad f'=\left\{1-\frac{(t+t')}{3f}\right\}\frac{1}{1/F-1/f}.$$

Hence the error in f' caused by neglecting the term in $(t+t')$ is about $\frac{100(t+t')}{3f}$ %. For instance, if $f=10{\cdot}0$ cm. and the thickness of the combination is $0{\cdot}8+0{\cdot}1$ cm., which are probable figures, the error is about 3 %; so the usual crude formula $\frac{1}{F}=\frac{1}{f}+\frac{1}{f'}$ should not be used if the determination has any pretensions to accuracy.

(iv) *Any number of lenses*

The results of §§ (ii) and (iii) can be extended to any number of lenses. For we can treat the combination of the first two lenses as a single lens, with principal points at L_1 and L_2; combining this with the third lens, the results of (ii) and (iii) are applicable to the combination of three lenses, and so on by successive steps.

In particular, this extension of (15) consists merely of the addition of a number of similar terms. One of these terms may correspond to a small air-gap; since $\mu = 1{\cdot}0$ for air, this air-gap will contribute nothing to $L_1 L_2$. But it is more usual to cement the component lenses together with Canada Balsam, for which $\mu = 1{\cdot}53$ approximately, so that where the glass lenses do not touch one another at their centres we shall have a Canada Balsam lens which will make its appropriate contribution to the terms on the right-hand side of (13). If then the refractive indices of all the lenses are nearly equal to 1·5, the distance between the principal points will be nearly *one-third of the over-all thickness of the combination.*

APPENDIX C

Bending of Beams

Let APB represent a beam, initially straight, whose cross-section and elasticity are uniform along its length, supported horizontally on two knife-edges at A and B; the weight of the beam is supposed to be negligibly small compared with the other forces in operation.

If a force F vertically downwards is applied to a point P, a from A and b from B, it is required to find the equations to the curves AP and BP.

For brevity, denote $a+b$ by l.

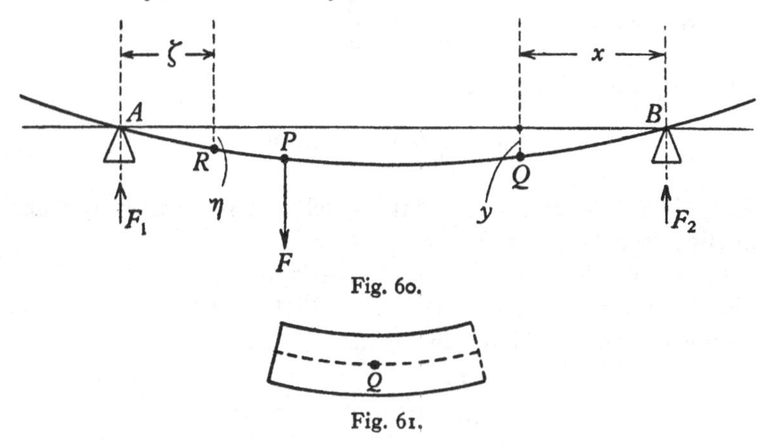

Fig. 60.

Fig. 61.

It is shown in text-books on Mechanics that if Fig. 61 represents a very short piece of a beam, initially straight, and the dotted line represents the 'neutral axis' (the line along which there is neither extension nor compression) passing through Q the C.G. of the cross-section, when the beam is slightly bent by forces having a bending moment M about Q, the dotted line will become curved, with a radius of curvature (r) equal to EI/M, where E is Young's modulus for the material and I is the moment of inertia of the cross-section of the beam about a horizontal axis in the plane of the cross-section and passing through Q.

In Fig. 60 this bending moment at Q is F_2x, so $r=EI/F_2x$. But taking moments about A for the forces on the beam, $F_2l=Fa$; hence

$$r=\frac{EIl}{aFx}. \qquad (1)$$

Now if the co-ordinates of Q on the curve PQB are x, y as shown in Fig. 60, it is shown in text-books on Differential Calculus that

$$r = -\left\{1 + \left(\frac{dy}{dx}\right)^2\right\}^{\frac{3}{2}} \bigg/ \frac{d^2y}{dx^2},$$

the sign being negative because the curve is concave to the axis of x.

For small deflections of the beam its inclination to the horizontal at any point (which is measured by dy/dx) is small, and we can therefore neglect $(dy/dx)^2$ in comparison with 1·0, and it follows that

$$r = -\frac{1}{\dfrac{d^2y}{dx^2}}.$$

So we get from (1) $$\frac{aFx}{EIl} = -\frac{d^2y}{dx^2}.$$

This equation is satisfied by

$$y = \frac{aF}{6EIl}(A' + Ax - x^3), \tag{2}$$

where A' and A are any constants, which may be verified by differentiating each side twice with respect to x.

Hence (2) is the equation to the curve PQB.

Similarly the equation to the curve PRA must be, if ζ, η are the co-ordinates of R, as shown in Fig. 60

$$\eta = \frac{bF}{6EIl}(B' + B\zeta - \zeta^3), \tag{3}$$

where B' and B are any constants.

We must now get the values of these four constants in this case. When $x=0$, $y=0$; so from (2),

$$A' = 0. \tag{4}$$

Similarly from (3) $$B' = 0. \tag{5}$$

And at P, whose $x=b$, let y, the vertical displacement of P caused by the force F, be denoted by n. Then from (2) and (4)

$$n = \frac{aF}{6EIl}(Ab - b^3).$$

Again, when $\zeta = a$, η will equal n, so from (3) and (5)

$$n = \frac{bF}{6EIl}(Ba - a^3).$$

Hence, subtracting,

$$(A-b^2)\frac{abF}{6EIl}=(B-a^2)\frac{abF}{6EIl} \quad \text{or} \quad A-B=b^2-a^2. \tag{6}$$

Again, at P, since the beam is continuous there, the tangents to the two curves ARP and BQP must form one straight line, so at that point $\frac{dy}{dx}=-\frac{d\eta}{d\zeta}$, the sign being negative because x and ζ are measured in opposite directions along AB.

But from (2) $\frac{dy}{dx}=\frac{aF}{6EIl}(A-3x^2)$, so at P it equals $\frac{aF}{6EIl}(A-3b^2)$.

Similarly $\frac{d\eta}{d\zeta}$ at P equals $\frac{bF}{6EIl}(B-3a^2)$.

Hence

$$a(A-3b^2)=-b(B-3a^2) \quad \text{or} \quad aA+bB=3ab(a+b). \tag{7}$$

Eliminating B from (6) and (7)

$$(a+b)A=3ab(a+b)+b(b-a)(b+a)=(a+b)(3ab+b^2-ab)$$

or $$A=b^2+2ab=b(l+a)=l^2-a^2.$$

Similarly $$B=l^2-b^2.$$

Hence, finally, substituting these values in (2) and (3), the equations for the two parts of the curve become

$$y=\frac{F}{6EIl}ax(l^2-a^2-x^2) \tag{8}$$

and $$\eta=\frac{F}{6EIl}b\zeta(l^2-b^2-\zeta^2). \tag{9}$$

In the particular case where the cross-section of the beam is a rectangle, of horizontal breadth c cm. and vertical thickness d cm., $I=\frac{1}{12}cd^3$; if F is caused by a load of W g., $F=Wg$, and y then becomes

$$\frac{2Wgax}{Ecd^3l}(l^2-a^2-x^2)$$

and η becomes $$\frac{2Wgb\zeta}{Ecd^3l}(l^2-b^2-\zeta^2).$$

INDEX

Printed in the United States
By Bookmasters